Selected Topics in Open Quantum Systems

Selected Topics in Open Quantum Systems

Edited by **Glenn Rogers**

NY RESEARCH PRESS
New York

Published by NY Research Press,
23 West, 55th Street, Suite 816,
New York, NY 10019, USA
www.nyresearchpress.com

Selected Topics in Open Quantum Systems
Edited by Glenn Rogers

International Standard Book Number: 978-1-63238-413-3 (Hardback)

Printed in the United States of America.

Contents

Preface

The main aim of this book is to educate learners and enhance their research focus by presenting diverse topics covering this vast field. This is an advanced book which compiles significant studies by distinguished experts in the area of analysis. This book addresses successive solutions to the challenges arising in the area of application, along with it; the book provides scope for future developments.

A quantum system in contact with its environment is called an open environment system. The aim of this book is to present certain aspects of contemporary theory on open quantum systems. It presents numerous updated topics, like modeling of quantum noise, detecting quantum entanglement, quantum communication procedures, computational intricacy in the evaluation of quantum operations etc. along with discussions on light propagation in optically dressed media, entropy and information measures for quantized electromagnetic fields media. The book aims to serve as a useful source of information for students as well as researchers interested in the extensive fields of open systems and quantum optics.

It was a great honour to edit this book, though there were challenges, as it involved a lot of communication and networking between me and the editorial team. However, the end result was this all-inclusive book covering diverse themes in the field.

Finally, it is important to acknowledge the efforts of the contributors for their excellent chapters, through which a wide variety of issues have been addressed. I would also like to thank my colleagues for their valuable feedback during the making of this book.

Editor

Section 1

Mathematical Modeling of Dynamical - Quantum Systems

Chapter 1

On Open Quantum Systems and Mathematical Modeling of Quantum Noise

Andrzej Jamiołkowski

Additional information is available at the end of the chapter

1. Introduction

This paper gives an introduction to some aspects of quantum processes described by quantum operations and quantum master equations. Quantum operations (quantum channels) and quantum master equations come into play whenever a mathematical description of irreversible time behaviour of quantum systems is investigated.

In modeling of physical systems for which time behaviour can be represented by stochastic processes one assumes that a system in question is described by certain mathematical model, for example, by random variables in classical case or by sets of non-commuting observables in quantum case, acting on an abstract probability space. In most cases our assumptions are not believed to be a fully realistic model of the physical reality, but nevertheless very often it turns out to be extremely useful.

In the algebraic formulation of quantum mechanics, a fixed quantum mechanical system is represented by an algebra $\mathcal{A}$ of operators acting on some Hilbert space $\mathcal{H}$. In this approach, the observables of the system are identified with self-adjoint elements in $\mathcal{A}$ and the physical states are given by positive unital functionals on $\mathcal{A}$. We will consider the case when $\mathcal{H}$ is finite-dimensional, then the set of states can be identified with the set of $\mathcal{S}(\mathcal{H})$ of density operators, that is, positive elements in $\mathcal{A}$ with unital trace. The evolutions of the system are described by transformations on the set $\mathcal{S}(\mathcal{H})$, or more generally, by linear maps between $\mathcal{S}(\mathcal{H})$ and $\mathcal{S}(\mathcal{K})$, where $\mathcal{H}$ and $\mathcal{K}$ represent two finite-dimensional Hilbert spaces.

Nonclassical correlations among subsystems of a composite quantum system, known as entanglement, can be precisely described mathematically but in a rather ineffective way. Here, by an effective way (effective procedure) we mean a method which uses only a finite number of arithmetic operations on elements of a given density matrix and allows us to formulate an answer to the question: is a given composite quantum system in a separable or entangled state?

It appears that the questions of the above type can be naturally connected with some problems of changes in the space of quantum states, that is with some aspects of quantum dynamics and properties of dynamical maps. Here, by dynamical maps we understand linear transformations that take one density operator to another. Moreover, this kind of connection between some linear transformations and states of composed quantum systems is a one-to-one type. In this context we will discuss the so-called Metzler operators which play a substantial role in quantum dynamics. By definition, an operator $\mathbb{K}$ acting on the space of hermitian matrices is said to be a Metzler operator if there exists a real number α_0 such that for all $\alpha > \alpha_0$ the resolvent of $\mathbb{K}$ is positive in the sense that

$$R(\alpha, \mathbb{K})\mathcal{P}(\mathcal{H}) \subseteq \mathcal{P}(\mathcal{H}),$$

where $\mathcal{P}(\mathcal{H})$ denotes the set of semipositive matrices.

In this paper we will discuss some effective methods of study of certain properties of quantum operations and we will use some methods and results which are typical for the research area known as *noncommutative Perron-Frobenius theory*. In particular, in our study of dynamical maps we will use the natural identifications

$$M_k(\mathbb{C}) \otimes B(\mathcal{H}) \cong M_k(B(\mathcal{H})) \cong B\left(\bigoplus_{i=1}^{k} \mathcal{H}\right),$$

where $\mathcal{H}$ denotes a finite-dimensional Hilbert space, $B(\mathcal{H})$ represents the set of all linear operators on $\mathcal{H}$ and symbols $M_k(\mathcal{A})$ for $k, 2, 3, \ldots$ denote $k \times k$ matrices with elements from an algebra $\mathcal{A}$. In our approach we will concentrate on a quantum analogy of the classical theory of positive maps also known as Perron-Frobenius theory. The Perron theorem on entrywise positive matrices and its generalization by Frobenius on entrywise nonnegative matrices have interested mathematicians since the results appeared at the beginning of last century. Later these theorems have been generalized to operators on a partially ordered real Banach spaces. This has motivated several authors to consider linear maps on a finite dimensional space which leave a fixed cone invariant.

Now, let us fix the notation. Let $\mathcal{H}$ denote the Hilbert space associated with a given quantum system $\mathcal{S}$. By $B(\mathcal{H})$ we will denote the set of all linear continuous operators on $\mathcal{H}$. Then the set of states of the system $\mathcal{S}$ is, by definition, represented by all semi-positive elements of $B(\mathcal{H})$ with trace equal to one. This set of states we will denote by $\mathcal{S}(\mathcal{H})$. The mentioned above connection between linear maps on $\mathcal{S}(\mathcal{H})$ and some operators constructed on the tensor product $\mathcal{H} \otimes \mathcal{H}$, is a one-to-one type. One can say that there is a one-to-one correspondence between properties of the maps

$$\Phi : B(\mathcal{H}) \rightarrow B(\mathcal{H}), \tag{1}$$

such that

$$\Phi(\mathcal{S}(\mathcal{H})) \subseteq \mathcal{S}(\mathcal{H}), \tag{2}$$

and properties of linear operators $W := J(\Phi)$ on $\mathcal{H} \otimes \mathcal{H}$, where J denotes this correspondence [1–4].

At first sight this seems strange that two entirely different issues, namely the inner structure and properties of operators on $\mathcal{H} \otimes \mathcal{H}$, and maps on the set of states are so strongly connected. But in fact there exists an intricate and strong link between them — there exists an isomorphism J between maps on $B(\mathcal{H})$ (these maps very often are called superoperators) and some properties of operators on $\mathcal{H} \otimes \mathcal{H}$ defined in an appropriate way. At present, various methods of study of entangled systems based on properties of positive maps are discussed in hundreds of papers and many monographs and textbooks (cf. [4–8]).

In the beginning of seventies it appeared that some natural questions connected with fundamentals of quantum mechanics (more precisely, with the theory of open quantum systems) lead to investigations of linear maps in a real Banach space of self-adjoint operators on a fixed Hilbert space [9, 32]. This concept of a Banach space with the partial order defined by a specific cone, namely, the cone of positive semidefinite operators, constitutes a basic idea in the description of open quantum systems.

In concrete applications one distinguishes two approaches to describe time evolutions (changes) of an open quantum system. One of them starts from a fixed physical model defined by a given Hamiltonian which determines the Schrödinger equation (von Neumann equation) or the master equation with a given generator of time evolution which is, in general, time dependent.

Summing up, as the fundamental objects in modern quantum theory one considers the set of states

$$S(\mathcal{H}) := \{\rho : \mathcal{H} \to \mathcal{H}; \quad \rho \geq 0, \quad \mathrm{Tr}\rho = 1\}, \tag{3}$$

and the set of bounded hermitean (self-adjoint) operators

$$\mathcal{B}_{\star}(\mathcal{H}) := \{Q : \mathcal{H} \to \mathcal{H}; \quad Q = Q^{\star}\}. \tag{4}$$

Time evolutions of systems are governed by linear master equations of the form (in the so-called Schrödinger picture)

$$\frac{d\rho(t)}{dt} = \mathbb{K}(t)\rho(t), \tag{5}$$

or in the dual form (in the so-called Heisenberg picture)

$$\frac{dQ(t)}{dt} = \mathbb{L}(t)Q(t), \tag{6}$$

where superoperators $\mathbb{K}$ and $\mathbb{L}$ act on operators from the sets $S(\mathcal{H})$ and $\mathcal{B}_{\star}(\mathcal{H})$, respectively. They represent dual forms of the same physical idea.

An alternative approach to the dynamics of an open quantum system relies on a stroboscopic picture and a discrete time evolution. We start from a mathematical construction of a quantum map on $\mathcal{S}(\mathcal{H})$ (in fact on $B_{*}(\mathcal{H})$) which is allowed by the general laws of quantum mechanics. Such approach is particulary useful if we want to investigate the set of all possible operators independently on whether the physical situation is exactly specified.

Both sets $\mathcal{S}(\mathcal{H})$ and $B_*(\mathcal{H})$ can be considered as subsets of the vector space $\mathcal{B}(\mathcal{H})$ of all bounded linear operators on $\mathcal{H}$ and they can be treated as scenes on which problems of quantum mechanical systems should be discussed. As in this paper we will consider finite-dimensional Hilbert spaces, so in fact $\mathcal{B}(\mathcal{H})$ denotes the set of *all* linear operators on $\mathcal{H}$, $\dim \mathcal{H} = d$. If we introduce the scalar product in $\mathcal{B}(\mathcal{H})$ by the equality

$$\langle A, B \rangle := \mathrm{Tr}\,(A^\star B)\,, \tag{7}$$

then $\mathcal{B}(\mathcal{H})$ can be regarded as yet another inner product space, namely the *Hilbert–Schmidt* space. It is not difficult to see that $\mathcal{B}_\star(\mathcal{H})$ with scalar product defined by (7) is a real vector space and $\dim \mathcal{B}(\mathcal{H}) = d^2$. It should be stressed that in this case we consider linear maps on $B(\mathcal{H})$ not only as maps on a vector space but also as maps on $B(\mathcal{H})$ equipped with a natural structure of an algebra. Such linear maps on the set of operators are called *superoperators* and their general form is well known. Namely, for a given superoperator Φ, $\Phi : B(\mathcal{H}) \to B(\mathcal{H})$, there always exists an operator-sum representation given by

$$\Phi(X) = \sum_{i=1}^{\kappa} A_i X B_i\,, \tag{8}$$

where A_i, B_i are elements of $B(\mathcal{H})$. A particular class of such maps, the so-called completely positive maps (or in physical terminology — *quantum operations*), plays a prominent role in formulation of *evolution* of open quantum systems and in the theory of *quantum measurements*. In the case of completely positive maps we have in the above formula $B_i = A_i^\star$ for all $i = 1, \ldots, \kappa$ (Kraus theorem). A comprehensive description of these problems from the physical point of view can be found in books [3, 14].

It is important to observe that most of the papers devoted to the classical Perron-Frobenius theory of nonnegative linear maps are concerned mainly with the existence problems. The purpose of this paper is, on the one hand, to connect the Frobenius theory of irreducible linear operators with the so-called Kraus representation of the completely positive linear maps and, on the other hand, to show that there exist some effective procedures which allow us to verify if a given quantum operation (i.e. a given completely positive map) is irreducible or not. Here by irreducibility we mean the natural generalization of this concept, introduced by Frobenius in his famous paper of 1912 year ([12]) and based on the specific block representation of nonnegative matrices, to a geometric approach formulated in terms of the invariance of faces of a fixed cone. In this context, by an effective procedure we understand a method which uses only a finite number of arithmetic operations on matrices K_i (Kraus coefficients) which define a fixed quantum operation (completely positive map)

$$\Phi(X) = \sum_{i=1}^{\kappa} K_i X K_i^\star\,. \tag{9}$$

Another important question may be formulated in the following way: for a fixed quantum operation (a superoperator) Φ defined by a set of Kraus operators $K_1, \ldots, K_\kappa$ there exists a decoherence-free subspace or not. By definition, a decoherence-free subspace (DFS) it is a subspace of the system's Hilbert space $\mathcal{H}$ which is invariant under non-unitary evolution. Alternatively formulated, DFS is this part of the system Hilbert space where the system is decoupled from the environment and thus its evolution is completely unitary. For a given quantum operation Φ defined by (9) DFS can exist or not. The questions is: can we check this property in an effective way or not?

The paper is organized as follows. In Section 2 we will review the concepts needed for our discussion based on geometric and operator theoretic formulation of the Perron-Frobenius theory and, on the other hand, we describe the structure of the cone of positive definite operators defined on a given Hilbert space. Moreover, some properties of faces of this cone are analyzed. Section 3 describes some properties of general positive maps (superoperators), and their representations as operators acting on doubled Hilbert space $\mathcal{H} \otimes \mathcal{H}$. The main results of the paper are contained in Sections 4 and 5. In Section 4 we formulate some effective methods of checking whether a given superoperator, i.e. a fixed quantum operation, is irreducible or not. A similar problem whether a given generator of time evolution in master equation description of evolution secures positivity of density operators or not, is discussed in Section 5.

2. Properties of cones

In order to start this section we recall some definitions and facts from the theory of cones and operators on a partially ordered vector space. A convex closed set K of a real normed space V is called a *wedge* iff

$$\alpha K + \beta K \subseteq K \tag{10}$$

for all $\alpha, \beta \geq 0$. A wedge K is called a *cone* if, in addition, we have

$$K \cap (-K) = \{\mathbf{0}\}, \tag{11}$$

where $\mathbf{0}$ denotes the zero element of V. If we have the equality $V = K - K$, then the cone K is called *generating* or *reproducing* (sometimes one also uses the term *full cone*).

If Φ is a linear transformation on V, $\Phi : V \to V$, then we denote by $r(\Phi)$ the *spectral radius* of Φ, i.e.

$$r(\Phi) := \max\{|\lambda| \, ; \, \lambda \in \sigma(\Phi)\}, \tag{12}$$

where $\sigma(\Phi)$ denotes the spectrum of Φ. For any cone K we let K° denote the interior of K and by ∂K we denote its boundary.

As is well known any fixed cone K in V determines a partial order in V. For this order we use the following terminology:

1. x is *nonnegative*, $x \geq 0$, iff $x \in K$;
2. x is *positive*, $x > 0$, iff $x \geq 0$ and $x \neq 0$;
3. x is *strictly positive*, $x \gg 0$, iff $x \in K^\circ$.

Now, let us define the concept of face which plays a basic role in the theory of irreducible operators. Let K be a cone in V. By a *face* F of K one understands a subset of K which is a cone and satisfies an extra condition: if $0 \leq y \leq x$ and $x \in F$, then y also belongs to F, $y \in F$.

Of course, if we fixed a basis in V, then we may regard vectors in V as column vectors in $\mathbb{R}^n$. In this case the positive orthant $\mathbb{R}^n_+$ constitutes a cone in $\mathbb{R}^n$ and exact description of $\partial\mathbb{R}^n_+$ is obvious, namely,

$$F_M := \{x \in \mathbb{R}^n_+; \quad x_i = 0 \text{ if } i \notin M\}, \tag{13}$$

where $M \subseteq \{1, 2, \ldots, n\}$.

If $E \subset K$, then we will denote by $\Omega(E)$ the intersection of all faces containing E. It is easily seen that $\Omega(E)$ is a face. It is called the face *generated by E*.

The set of all operators $\Phi : V \to V$ such that $\Phi(K) \subseteq K$ we will denote by $\Pi(K)$. Let $B(V)$ be the set of all linear operators on V. Then we have

$$\Pi(K) := \{\Phi; \Phi(K) \subseteq K\} \subseteq B(V) \tag{14}$$

and $\Pi(K)$ is a cone in $B(V)$. The elements of the set $\Pi(K)$ are said to be *K-nonnegative operators*. In particular, the operator Φ on V is called *K-positive* in case $\Phi(K \setminus \{0\}) \subseteq K^o$. The set of all K-positive maps will be denoted by $\Pi^+(K)$.

Now we introduce one of the main ideas of the Perron-Frobenius theory both in commutative and non-commutative case. For a fixed K in V a natural generalization of the concept of an irreducible matrix is the following: Φ is *K-irreducible* if and only if Φ leaves invariant no face of K except $\{0\}$ and K itself and κ describes the minimal length of Φ. In other words, an operator in $\Pi(K)$ is K-reducible iff it leaves invariant a nontrivial face of K.

Another, strictly equivalent, definition of K-irreducibility can be given by the following theorem: An operator $\Phi \in \Pi(K)$ is K-irreducible if and only if no eigenvector of Φ lies on the boundary of K. In fact, one can say even more: An operator $\Phi \in \Pi(K)$ is K-irreducible if and only if Φ has exactly one (up to scalar multiples) eigenvector in K and this vector belongs to K^o. Moreover, for any proper cone K we have

$$\Pi^+(K) \subseteq \widetilde{\Pi}(K) \subseteq \Pi(K), \tag{15}$$

where $\widetilde{\Pi}(K)$ denotes the set of all K-irreducible operators. If $K = \mathbb{R}^n_+$, then the both definitions, Frobenius one and the above, coincide. For details, see e.g. [13, 15]

Some important spectral properties of K-nonnegative operators are summarized in the following theorems, which one can consider as natural generalizations of well-known classical results.

Theorem 1. *Let $\Phi \in \Pi^+(K)$. Then we have*

a) *the spectral radius of the operator Φ is a simple eigenvalue of Φ, greater than the magnitude of any other eigenvalue;*

b) *an eigenvector of Φ corresponding to $r(\Phi)$ belongs to K^o;*

c) *no other eigenvector of Φ (up to scalar multiples) belongs to K.*

Theorem 2. *Let $\Phi \in \Pi(K)$. Then the following hold*

a) *$r(\Phi)$ is an eigenvalue of Φ ;*

b) *K contains an eigenvector of Φ corresponding to $r(\Phi)$;*

c) *if $\Phi \leq \Psi$, then $r(\Phi) \leq r(\Psi)$.*

Theorem 3. *Let $\Phi \in \widetilde{\Pi}(K)$. Then the following hold*

a) *$r(\Phi)$ is a simple eigenvalue of Φ;*

b) *no eigenvector of Φ lies on the boundary of K ;*

c) *Φ has exactly one (up to scalar multiples) eigenvector in K and this vector belongs to K^o;*

d) *$(I + \Phi)^{n-1} \in \Pi^+(K)$, where $n = \dim V$.*

For proofs of the above theorems consult [13, 17].

We will conclude this section with some comments on the two special cases which are important from the point of view of physics, namely, $K = \mathbb{R}^d_+$ and $K = M^+_d(\mathbb{C})$. Here, the last symbol denotes the set of all semipositive operators on the space $\mathbb{C}^d$, and $\mathbb{C}^d$ is regarded as a representation of a d-dimensional Hilbert space. For the case $K = \mathbb{R}^d_+$, the whole story reduces to the classical Perron-Frobenius theory and in statistical physics we use this theory for the so-called mesoscopic description of classical systems.

The case $K = M^+_d(\mathbb{C})$ plays a fundamental role in the description of representations of states for open quantum systems. In this case, $B(\mathcal{H})$ is a d^2-dimensional vector space. According to (4) we denote by $B_*(\mathcal{H})$ the set of all self-adjoint operators on $\mathcal{H}$ which can be naturally considered as a d^2-dimensional *real* Banach space. At the same time, $B(\mathcal{H})$ can be regarded as a Hilbert space with the scalar product defined by (7). The vector space $B_*(\mathcal{H})$ of all Hermitian (self-adjoint) operators on $\mathcal{H}$ constitutes an d^2-dimensional, real subspace of the Hilbert-Schmidt space. One can use the "internal structure" of vectors in $B_*(\mathcal{H})$ to define a positive cone. By definition, a *semipositive* element of $B_*(\mathcal{H})$ is an operator A on $\mathcal{H}$ such that $\langle\psi|A|\psi\rangle$ is real and nonnegative for all vectors $|\psi\rangle$ from $\mathcal{H}$. Of course, one can equivalently define a positive element of $B(\mathcal{H})$ as a self-adjoint operator with nonnegative eigenvalues. The set of all semipositive operators on $\mathcal{H}$ we will denote by $B^+_*(\mathcal{H})$ or $\mathcal{P}(\mathcal{H})$. In particular, if we have the inequality $\langle\psi|A|\psi\rangle > 0$ for all $|\psi\rangle$ from $\mathcal{H}$, then we say that A is *positive*.

Let $\mathcal{P}_n$ denote the set of all orthogonal projections, i.e. $A \in \mathcal{P}_n$ if and only if $A \in B_*(\mathcal{H})$ and $A^2 = A$. With the natural order on projections, namely, $A \leq B$ iff $\mathrm{Im}(A) \subseteq \mathrm{Im}(B)$, the mapping from $A \in \mathcal{P}_n$ to $F(\mathcal{P}(\mathcal{H}))$, where $F(K)$ denotes the set of all faces of K, is an order preserving isomorphism.

Now, we formulate an important characterization of all faces of any cone $\mathcal{P}(\mathcal{H})$. It appears that each face of $\mathcal{P}(\mathcal{H})_d$, where the suffix d denotes $\dim \mathcal{H}$, is isomorphic to $\mathcal{P}(\mathcal{H})_m$ for some m, where $0 \leq m \leq d$.

Theorem 4. *If $B \in \mathcal{P}(\mathcal{H})$ is of rank m, then there exists a unitary U such that*

$$\Omega(B) = U^*(\mathcal{P}(\mathcal{H})_m \oplus 0_{d-m})U. \tag{16}$$

Conversely, if U denotes a unitary operator on $\mathcal{H}$, then $U^*(\mathcal{P}(\mathcal{H})_m \oplus 0_{d-m})U$ for $m = 0,1,...,d$ are faces of $\mathcal{P}(\mathcal{H})_d$.

It is well known that if K is a polyhedral cone, then for all faces of K we have

$$\operatorname{span} F + \operatorname{span} F^{\triangleleft} = V, \tag{17}$$

where $F^{\triangleleft}$ denotes the so-called complementary face of F defined by

$$F^{\triangleleft} := \{z \in K^*; \langle z, x \rangle = 0 \text{ for all } x \in F\} \tag{18}$$

and V denotes the ambient space of K. The *residual subspace* of F is meant to measure "to what extent F is nonpolyhedral". It is defined as

$$\operatorname{res}(F) := (\operatorname{span} F + \operatorname{span} F^{\triangleleft})^{\perp}. \tag{19}$$

A list of several examples of cones, along with the description of their faces and residual subspaces, is contained in [18].

3. Positive maps on $B_*(\mathcal{H})$

It is well known that if a linear map $\Phi : B(\mathcal{H}) \to B(\mathcal{H})$ sends the set $B_*(\mathcal{H})$ of all hermitian elements of $B(\mathcal{H})$ into itself, then Φ can be represented in the form

$$\Phi(X) = \sum_{i=1}^{\kappa} a_i A_i X A_i^*, \tag{20}$$

where $A_i \in B(\mathcal{H})$, and a_i for $i = 1,2,...,\kappa$ are real numbers (cf. eg. [3,19]). In general, all maps of the above form are hermitian-preserving. However, the representation (20) is not unique. In general, for a given Φ , there exist many possible representations of the form (20). For a given Φ the smallest κ in (20) is called the *minimal length* of Φ and this minimal length is always smaller or equal to $(\dim \mathcal{H})^2$. If we assume that the operators A_i for $i = 1,2,\ldots,\kappa$ are linearly independent, then κ in (20) must be minimal.

According to the general definition introduced in Section 2 a *positive map* Φ is a linear map from $B(\mathcal{H})$ into itself, which leaves $\mathcal{P}(\mathcal{H})$ invariant. Now, Φ is called *k-positive* if its *k-amplification* $\Phi_{(k)} := \mathbb{I}_k \otimes \Phi$ that is the map

$$\mathbb{I}_k \otimes \Phi : M_k(\mathbb{C}) \otimes B(\mathcal{H}) \to M_k(\mathbb{C}) \otimes B(\mathcal{H}) \tag{21}$$

is positive. Here $M_k(\mathbb{C})$ denotes as usual the set of all $k \times k$ complex matrices. It is not difficult to observe that we can identify the set $M_k(\mathbb{C}) \otimes \mathcal{A}$, where for simplicity we denote the algebra $B(\mathcal{H})$ by $\mathcal{A}$, with the set of all $k \times k$ matrices $M_k(\mathcal{A})$ with entries from $\mathcal{A}$.

The map Φ is called completely positive if it is k-positive for all $k = 1, 2, \ldots$. This terminology goes back to Stinespring [20], cf. also [4]. It is well known that for d-dimensional Hilbert space $\mathcal{H}$, d-positive maps on $B(\mathcal{H})$ are already completely positive [21].

Let us observe that all hermitian-preserving maps which are not only positive but completely positive can be written in the form (20) with positive a_i, $i = 1, \ldots, \kappa$, i.e. by

$$\Phi(X) = \sum_{i=1}^{\kappa} K_i X K_i^{\star}, \tag{22}$$

where $K_i := \sqrt{a_i} A_i$, and $\kappa \leq d^2$. The above expression is called the Kraus representation of a completely positive map Φ and, in case of the finite-dimensional Hilbert space $\mathcal{H}$, can be regarded also as a definition of the completely positive map. This representation is very useful in quantum information theory. In particular, completely positive maps are used to describe all quantum operations, quantum channels and to model quantum devices.

Let us observe that an equivalent representation of the evolution described by expression (22) can be made in terms of the operator W mentioned in Section 1. This operator is connected in a one-to-one way with the superoperator Φ (quantum map on $\mathcal{S}(\mathcal{H})$) by the formula

$$W(\Phi) := (\mathbb{I} \otimes \Phi) \left(|\psi_+\rangle\langle\psi_+| \right), \tag{23}$$

where

$$|\psi_+\rangle := \sum_{i=1}^{d} |e_i\rangle \otimes |e_i\rangle, \tag{24}$$

and $\{|e_i\rangle\}$ denotes a basis in $\mathcal{H}$. The operator W, acting on the doubled space $\mathcal{H} \otimes \mathcal{H}$ is, after normalization, called J-state and the correspondence defined by (23) is called J-isomorphism. According to the best knowledge of the author this relationship between Φ and $W(\Phi)$ was applied to problems of evolution of quantum systems for the first time in papers cited in [1] and [2] in the begining of seventies. More details about definition (23) is given in the next section.

Three decades later it was shown by R. Timoney [22] that a positive map Φ (positive superoperator) which is m-positive, where $m = \lfloor \sqrt{\kappa} \rfloor$ must be completely positive. Here $\lfloor \sqrt{\kappa} \rfloor$ denotes the integer part of the number $\sqrt{\kappa}$. In other words, if a positive map $\Phi : B(\mathcal{H}) \to B(\mathcal{H})$, $\dim \mathcal{H} = d$, has a minimal length κ and Φ is m-positive, for some $m < n$ such that $(m+1)^2 > \kappa$, then Φ is already completely positive (cf. also [19]).

Now, let us observe that one can apply the results stated in Theorems 1 – 3 from Section 2, to the particular cone $\mathcal{P}(\mathcal{H})$ in $B(\mathcal{H})$, $\dim \mathcal{H} = d$. In particular, Theorem 3 describes properties of irreducible superoperators and according to this theorem we have: for irreducible Φ, the spectral radius of Φ is a simple eigenvalue of the superoperator and Φ has exactly one (up to scalar coefficient) eigenvector in $\mathcal{P}(\mathcal{H})$ and this vector belongs to $\mathcal{P}(\mathcal{H}) \setminus \partial\mathcal{P}(\mathcal{H})$. One can say even more. Let $\mathcal{P}_d$ denote the set of all orthogonal projections, i.e. $A \in \mathcal{P}_d$ iff $A^2 = A$ and $A = A^*$. Then we have [23]

Theorem 5. *The following statements are equivalent for a positive map on $\mathcal{P}(\mathcal{H})$.*

1.) *There is a nontrivial (that is different from $\{0\}$ and $\mathcal{P}(\mathcal{H})$) face of $\mathcal{P}(\mathcal{H})$ that is invariant under Φ;*

2.) *There is nontrivial projection $P \in \mathcal{P}_d$ and a positive real number $\lambda > 0$ such that $\Phi(P) \leq \lambda P$;*

3.) *There is a nontrivial projection $P \in \mathcal{P}_d$ such that the subalgebra $P(\mathcal{P}(\mathcal{H}))P$ is invariant under Φ.*

In order to produce nontrivial examples of irreducible positive maps and in certain cases to characterize all irreducible maps within the class of completely positive maps we will use the following consequences of the Kraus representations of completely positive maps.

A family of closed subspaces of a given Hilbert space is called trivial if this family contains only $\{0\}$ and $\mathcal{H}$. For a fixed operator $X \in B(\mathcal{H})$ we will denote by $\mathrm{Inv}\,(X)$ the set of all invariant subspaces of X. Now, we can state the following theorem which is a reformulation of some results from [23].

Theorem 6 (Farenick). *Let Φ denote a superoperator on $B(\mathcal{H})$ which is $\mathcal{P}(\mathcal{H})$-positive. If Φ is completely positive, then there exist some operators $A_1, \ldots, A_\kappa$ such that $\Phi(X) = \sum_j A_j X A_j^*$. Completely positive Φ is irreducible if and only if the Kraus operators A_j do not have a nontrivial common invariant subspace in $\mathcal{H}$.*

To better understand the above theorem, let us observe that if Kraus operators do not have a common invariant subspace, i.e., are such that $\cap_j \mathrm{Inv}(A_j)$ is trivial and $\Phi(P) \leq \lambda P$ for some $\lambda \geq 0$ and $P \in \mathcal{P}_n$, then we have

$$\langle \Phi(P)\psi|\psi\rangle = \sum_{j=1}^{\kappa} \langle PA_j\psi|A_j\psi\rangle \leq \lambda\langle P\psi|\psi\rangle \,. \tag{25}$$

The left-hand side of the above equality is nonnegative for all $\psi \in \mathcal{H}$. On the other hand for $\psi \in \ker P$, we have $\langle P\psi|\psi\rangle = 0$ on the right-hand side. In this way the equality (25) implies that $\langle PA_j\psi|A_j\psi\rangle = 0$ for each $j = 1, \ldots, \kappa$. This means that $\langle PA_j\psi|PA_j\psi\rangle = 0$, if we remember that $P^2 = P$, $P^* = P$. In consequence, $\ker P \in \cap_j \mathrm{Inv}(A_j)$, that is $\ker P$ is either $\{0\}$ or $\mathcal{H}$.

4. Some effective methods in the study of positive maps

If we compare the set of all linear maps $\Phi : B(\mathcal{H}) \to B(\mathcal{H})$ and the set of all linear operators on $\mathcal{H} \otimes \mathcal{H}$, i.e. $\mathcal{B}(\mathcal{H} \otimes \mathcal{H})$, where $\mathcal{H}$ is a d-dimensional Hilbert space, then it is easy to see that these sets represent isomorphic spaces (as two vector spaces with the same dimension). Because of linearity of maps Φ, any fixed map is fully defined if we know the values of Φ on elements of arbitrary basis of the space $B(\mathcal{H})$. For example, we know Φ if we know the values of Φ on elements of $B(\mathcal{H})$ of the form $|e_i\rangle\langle e_j|$

$$\Phi(E_{ij}) := \Phi(|e_i\rangle\langle e_j|)\,, \tag{26}$$

where $|e_i\rangle$, for $i = 1,\ldots,d$ are orthonormal vectors in $\mathcal{H}$. Usually, we assume that this is the so-called natural zero-one basis in $\mathbb{C}^d$; $|e_i\rangle$ represents vector with 1 on i-th position and zeros on other places. Let us observe that operators of the form $|e_i\rangle\langle e_j|$ which act on vectors from $\mathbb{C}^d$ by the rule

$$(|e_i\rangle\langle e_j|)|\psi\rangle = \langle e_j|\psi\rangle|e_i\rangle \tag{27}$$

constitute a basis in the space $B(\mathcal{H})$ and are projectors in directions $|e_i\rangle$.

Now, an important question is: how to relate, for a given map $\Phi : B(\mathcal{H}) \to B(\mathcal{H})$, an operator on $\mathcal{H}\otimes\mathcal{H}$ (that is an element from $\mathcal{B}(\mathcal{H}\otimes\mathcal{H})$) in such a way in order to have a representation of Φ which will be useful for description of properties of this map. In other words, we look for a specific isomorphism between maps Φ on $B(\mathcal{H})$ and elements from $\mathcal{B}(\mathcal{H}\otimes\mathcal{H})$. It appears that this isomorphism can be defined in the following way: we take values $\Phi(E_{ij})$ from (26) and define an element $W(\Phi)$ from $\mathcal{B}(\mathcal{H}\otimes\mathcal{H}) \equiv B(\mathcal{H})\otimes B(\mathcal{H})$ by the formula

$$W(\Phi) := \sum_{i,j=1}^{d} E_{ij}\otimes\Phi(E_{ij}) \tag{28}$$

which is equivalent to (23).

Now, it can be shown that Φ preserves the cone $\mathcal{P}(\mathcal{H})$ of nonnegative operators if and only if [1,2]

$$\langle\varphi|\otimes\langle\psi|W(\Phi)|\varphi\rangle\otimes|\psi\rangle \geq 0 \tag{29}$$

for all $|\varphi\rangle$, $|\psi\rangle$ from $\mathcal{H}$, and Φ is completely positive if and only if operator $W(\Phi)$ is semipositive on $\mathcal{H}\otimes\mathcal{H}$. We will also use this representation in Section 5, to discuss some properties of master equations.

On the other hand, we know that any completely positive map (superoperator) on $B(\mathcal{H})$ can be represented by a set of Kraus operators $K_1,\ldots,K_\kappa$ and the superoperator Φ is irreducible if the operators K_j ($j = 1,\ldots,\kappa$) do not have a nontrivial common invariant subspace in the space $\mathcal{H}$. The question connected with the existence of DFS with a fixed dimension can be also related to the following problem: is it possible to verify whether operators $K_1,\ldots,K_\kappa$ have — or do not have — a common invariant subspace of dimension $m < d$, by an effective procedure? For $m = 1$ an answer to this question was given by Dan Shemesh in 1984 [24].

Theorem 7. *Let K_1 and K_2 denote two matrices acting on $\mathcal{H} = \mathbb{C}^d$. A common eigenvector of K_1 and K_2 exists if and only if the subspace defined by*

$$\mathcal{M}_1 := \bigcap_{\alpha,\beta}^{d-1} \ker\left[K_1^\alpha,\, K_2^\beta\right] \tag{30}$$

is nontrivial, that is $\mathcal{M}_1 \neq \{\mathbf{O}\}$ or, in other words, $\dim\mathcal{M}_1 > 0$. Here the symbol $[\cdot,\cdot]$ denotes the commutator of the matrices and $\alpha,\beta \in [1,\ldots,d-1]$.

The above theorem is connected in natural way with the concept of partially commutative operators. Two operators K_1 and K_2 (complex matrices $d \times d$) which do not commute, $[K_1, K_2] \neq 0$, are said to be partially commuting if K_1 and K_2 have a common invariant subspace (at least one common eigenvector). The reason for introducing this term is obvious: if $|x\rangle \in \mathcal{H}$ is a nonzero vector such that

$$K_1|x\rangle = \lambda_1|x\rangle \quad \text{and} \quad K_2|x\rangle = \lambda_2|x\rangle, \tag{31}$$

then there exists a nontrivial common invariant subspace of K_1, K_2 on which these operators (matrices) commute.

As it was stressed in [25], the genuine meaning of the subspace $\mathcal{M}_1$, can be stated as follows.

Theorem 8. *A subspace $\mathcal{M}_1$ is invariant with respect to both matrices K_1 and K_2, and moreover, K_1 and K_2 commute on $\mathcal{M}_1$. Every subspace of $\mathcal{H}$, which is invariant under K_1 and K_2 and on which K_1 and K_2 commute is contained in $\mathcal{M}_1$.*

The condition of Theorem 8, that is $\dim \mathcal{M}_1 > 0$, can be formulated in a constructive form. To this end let us define the matrix

$$\mathcal{O} := \sum_{\alpha,\beta}^{d-1} \left[K_1^\alpha, K_2^\beta\right]^* \left[K_1^\alpha, K_2^\beta\right]. \tag{32}$$

The matrices K_1 and K_2 have common eigenvectors if and only if the matrix $\mathcal{O}$ is singular, i.e. $\det \mathcal{O} = 0$.

It is not difficult to check that if one of the operators K_1, K_2 has distinct eigenvalues (let say K_1 has this property), then the last expression for $\mathcal{O}$ reduces to

$$\mathcal{O} := \sum_{\alpha}^{d-1} \left[K_1^\alpha, K_2\right]^* \left[K_1^\alpha, K_2\right], \tag{33}$$

and the condition $\det \mathcal{O} = 0$ is simplified from the point of view of calculations.

One can say even more. If one of operators K_j, $j = 1, \ldots, \kappa$, has distinct eigenvalues (once again, let say K_1 has this property) then the appropriate operator (matrix) $\mathcal{O}$ takes the form

$$\mathcal{O} := \sum_{\alpha=1}^{d-1} \sum_{j=2}^{\kappa} \left[K_1^\alpha, K_j\right]^* \left[K_1^\alpha, K_j\right], \tag{34}$$

and the condition $\det \mathcal{O} = 0$ tells us that operators $K_1, \ldots, K_\kappa$ have a common eigenvector (Jamiołkowski, in preparation).

Now, using the concept of the so-called standard polynomials and the Amitsur-Levitzki theorem [27,28], we can generalize the criterion of Shemesh in another way. Recall that the standard polynomial of degree n is the polynomial in non-commuting variables $X_1, \ldots, X_n$ of the form

$$S_r(X_1, \ldots, X_n) := \sum \operatorname{sign}(\sigma) X_{\sigma(1)} \cdots X_{\sigma(n)}. \tag{35}$$

The summation here is assumed over all permutations of $1, \ldots, n$. The importance of the standard polynomials is underlined by the following Amitsur-Levitzki theorem.

Theorem 9. *The full matrix algebra $\mathbb{M}_d(\mathbb{C})$ satisfies the standard identity $S_{2d}(X_1, \ldots, X_{2d}) \equiv 0$. Moreover, the algebra $\mathbb{M}_d(\mathbb{C})$ does not satisfy any polynomial of degree less than $2d$.*

Let us observe that according to the above theorem, the algebra $\mathbb{M}_{d+1}(\mathbb{C})$ cannot satisfy the identity for $n = 2d$. In other words, the algebra $\mathbb{M}_k(\mathbb{C})$ satisfies the identity $S_{2d} = 0$ for $k \leq d$ and does not satisfy this identity for $k \geq d+1$.

Now, we are ready to discuss a generalization of the Shemesh theorem. Namely, we introduce the family of the subspaces

$$\mathcal{M}_k := \bigcap \ker \left[S_{2k}(N_1, \ldots, N_{2k}) N_{2k+1}\right], \tag{36}$$

where S_{2k} denotes the standard polynomial of degree $2k$ and the intersection is taken over all $(2k+1)$-tuples of matrices from the algebra $\mathcal{A}$ generated by two elements (Kraus operators K_1 and K_2). One can prove:

Theorem 10. *Assume that $\mathcal{M}_k$ satisfies $\dim \mathcal{M}_k > 0$. Then $\mathcal{M}_k$ is an invariant subspace of $\mathcal{A}$ and elements of this algebra restricted to $\mathcal{M}_k$ satisfy the identity $S_{2k} \equiv 0$; i.e. for all $N_1, \ldots, N_{2k}$ from $\mathcal{A}$ and $X \in \mathcal{M}_k$ we have*

$$S_{2k}(N_1, \ldots, N_{2k}) X = 0. \tag{37}$$

It is not obvious from (36) that $\mathcal{M}_k$ can be constructed by an effective procedure; however, it is possible to show that these subspaces can be constructed by a finite number of arithmetic operations [28].

As a conclusion of this section we can say that if in the Kraus representation of any fixed completely positive map Φ,

$$\Phi(X) = \sum_{i=1}^{K} K_i X K_i^*, \tag{38}$$

at least two Kraus operators do not have a common eigenvector, then the map Φ is irreducible. Moreover, using the generalization of Shemesh's theorem we can check in an effective way whether the algebra $\mathcal{A}$ generated by Kraus operators has a decoherence-free subspace of dimension $m \geq 2$, or not. In particular, according to (34), if one of Kraus' operators has distinct eigenvalues, the condition that $\det \mathcal{O} = 0$ is simplified from the point of view of efficiency of calculations.

5. Master equations and Metzler operators

In quantum physics, quantum chemistry and related fields by master equation one understands linear differential equation of the form

$$\frac{d\rho(t)}{dt} = \mathbb{K}(t)\rho(t) \tag{39}$$

with the following property: if $\rho(t_0)$ belongs to $\mathcal{P}(\mathcal{H}) = B_*^+(\mathcal{H})$ and satisfies the condition $\mathrm{Tr}(\rho(t_0)) = 1$, that is $\rho(t_0) \in \mathcal{S}(\mathcal{H})$, then the trajectory emanating from $\rho(t_0)$ stays in $\mathcal{S}(\mathcal{H})$ for all $t \geq t_0$. Here by trajectory we understand the solution $t \to \rho(t)$ to equation (39) with initial condition $\rho(t_0)$. In other words, a question of considerable physical, as well as, theoretical interest is the following. Under what conditions on $\mathbb{K}(t)$ does every solution of (39) which originates in $\mathcal{S}(\mathcal{H})$ remains in $\mathcal{S}(\mathcal{H})$ for all $t \geq 0$. In order to formulate the answer to this question in general form we say a few words on general solutions of equations in $B_*(\mathcal{H})$ with time dependent generators.

Let $\mathcal{D} \subset \mathbb{R}^1$ be an interval and let $\mathbb{K} : \mathcal{D} \to B_*(\mathcal{H})$ be a continuous operator-valued function with domain $\mathcal{D}$. Consider the linear differential equation

$$\frac{d\gamma(t)}{dt} = \mathbb{K}(t)\gamma(t)\,. \tag{40}$$

It is well known that for each $\gamma(t_0)$, equation (40) has a unique solution $\mathcal{D} \ni t \to \gamma(t) \in B_*(\mathcal{H})$ which is dependent in a linear way on $\gamma(t_0)$. Therefore for each pair t, t_0 from $\mathcal{D}$ one can define a linear operator $\Phi(t, t_0)$, $t \geq t_0$, by the formula

$$\Phi(t,t_0)\gamma(t_0) := \gamma(t,t_0,\gamma(t_0))\,, \tag{41}$$

where $\gamma(t, t_0, \gamma(t_0))$ satisfies (40), with the initial condition $\gamma(t_0)$.

The operator $\Phi(t, t_0)$ is usually called the *evolution operator* or the *propagator* of the time evolution defined in $B_*(\mathcal{H})$ by the generator $\mathbb{K}(t)$. The following properties characterize the propagator $\Phi(t, t_0)$:

1) $\Phi(t,u) \cdot \Phi(u,s) = \Phi(t,s)$ for all $t,u,s \in \mathcal{D}$,
2) $\Phi(t,s)$ is differentiable with respect to t and

$$\frac{d\Phi(t,s)}{dt} = \mathbb{K}(t)\Phi(t,s)\,, \tag{42}$$

3) $\Phi(t,t) = \mathbb{I}$ for all $t \in \mathcal{D}$, where $\mathbb{I}$ denotes the identity operator,
4) if $\mathbb{K}(t) = \mathbb{K}$ for all $t \in \mathcal{D}$, then

$$\Phi(t,t_0) = \exp\{(t-t_0)\mathbb{K}\} = \sum_{m=0}^{d-1} \alpha_m(t-t_0)\mathbb{K}^m\,,$$

where $\alpha_m(t - t_0)$ for $m = 0, \ldots, d-1$ are some analytic functions defined by the structure of $\mathbb{K}$ (cf. eg. [29]).

We say that $\mathbb{K}(t)$ defines a positive evolution if

$$\Phi(t, t_0) \geq 0 \tag{43}$$

for all $t \geq t_0,\ t, t_0 \in \mathcal{D}$. Here, the positivity of $\Phi(t, t_0)$ is understood with respect to the cone $\mathcal{P}(\mathcal{H}) = B_*^+(\mathcal{H})$.

An operator $\mathbb{K}$ (superoperator with respect to elements of $B_*(\mathcal{H})$) is characterized by its spectrum $\sigma(\mathbb{K})$ and the resolvent set $\omega(\mathbb{K})$. Now, a closed operator $\mathbb{K}$ is said to be a Metzler operator if there exists a real number α_0 such that for all $\alpha > \alpha_0$ the resolvent of the operator $\mathbb{K}$ is positive, that is,

$$R(\alpha, \mathbb{K})\mathcal{P}(\mathcal{H}) \subseteq \mathcal{P}(\mathcal{H})\,. \tag{44}$$

One uses this terminology since Metzler operators are straightforward generalization of Metzler matrices. Indeed, let us take the space $\mathbb{R}^d$ and a cone $\mathbb{R}_+^d$. Suppose $\mathbb{K} \in M_d(\mathbb{R})$ is a Metzler matrix, that is there exists $b \in \mathbb{R}^1$ such that $\mathbb{K} + b\mathbb{I}$ is positive, which means that all entries of the matrix are nonnegative real numbers.

One can say immediately that for $\alpha > r(\mathbb{K} + b\mathbb{I})$ the resolvent

$$R(\alpha, \mathbb{K} + b\mathbb{I}) = (a\mathbb{I} - (\mathbb{K} + b\mathbb{I}))^{-1} = [(a - b)\mathbb{I} - \mathbb{K}]^{-1} \tag{45}$$

is a positive operator, that is according to the above definition, Metzler matrix represents a Metzler operator. Vice versa, if $\mathbb{K}$ is a Metzler operator, the off-diagonal elements of $\mathbb{K} \in M_d(R)$ are all nonnegative, that is, $\mathbb{K}$ is a Metzler matrix.

It follows directly from our definition of Metzler operators that a fixed closed operator $\mathbb{K}$ is a Metzler operator if and only if also $\mathbb{K} + b\mathbb{I}$ is a Metzler operator for some $b \in \mathbb{R}^1$. Now, using the Neumann representation of $R(\alpha, \mathbb{K})$ for large $\alpha > 0$, one can see that every operator on finite-dimensional $B_*(\mathcal{H})$ for which there exists $b \geq 0$ such that $\mathbb{K} + b\mathbb{I}$ is positive, is a Metzler operator on $B(\mathcal{H})$.

It was proved by Elsner [30] that the following conditions for an operator $\mathbb{K}$ on any finite-dimensional Banach space V ordered by a closed, solid, convex cone C are equivalent:

1) $\mathbb{K}$ is a Metzler operator,
2) $\exp(t\mathbb{K})$ is positive for all $t \geq 0$,
3) $|x\rangle \in C, \langle v| \in C^*, \langle v|x\rangle = 0 \Rightarrow \langle v|\mathbb{K}x\rangle \geq 0$.

The condition 2) is often called exponential positivity or exponential nonnegativity. From 3) one can see that the Metzler operators constitute a convex cone. However, this cone in not pointed — it contains all scalar multiples of the identity operator. In fact we can say that the set of all Metzler operators constitutes a wedge.

Now, let us return to our system $B_*(\mathcal{H})$ and cones $\mathcal{P}(\mathcal{H}) = B_*^+(\mathcal{H})$. As we know, in this case $\rho(t) \in \mathcal{P}(\mathcal{H})$ for all $t \geq 0$ if

$$\langle \psi| \otimes \langle \varphi|W(\mathbb{K})|\psi\rangle \otimes |\varphi\rangle \geq 0 \tag{46}$$

([1] and [2], cf. also [3]). This means that the last inequality (the so-called block positivity [3]) is a sufficient condition for preservation of positivity by $\Phi(t, t_0) = \exp\left[(t - t_0)\mathbb{K}\right]$. But in fact the condition (46) is to strong — it is a sufficient condition but not a necessary. It is enough to assume that $\mathbb{K}$ satisfies the weaker condition — $\mathbb{K}$ should be a Metzler operator.

It appears that one can extend the conditions discussed above to the time dependent operator $\mathbb{K}(t)$. Namely, we can state that the operator family $\mathbb{K}(t)$ generates a positive evolution of the interval $\mathcal{D}$ if and only if $\mathbb{K}(t)$ is Metzler operator for all $t \in \mathcal{D}$ (cf. [31]).

It is important to observe that the property of being a Metzler operator can be expressed using the defined in (28) isomorphism $W = J(\mathbb{K})$. Namely, an operator $\mathbb{K}$ is a Metzler operator if and only if for all $|\psi\rangle$, $|\varphi\rangle$ such that they are orthogonal, we have inequality

$$\langle \psi| \otimes \langle \varphi| W(\mathbb{K}) |\psi\rangle \otimes |\varphi\rangle \geq 0\,, \qquad |\psi\rangle \perp |\varphi\rangle\,. \tag{47}$$

Indeed, as we know the superoperator $\mathbb{K}$ is a Metzler operator if $\langle A|\mathbb{K}(B)\rangle \geq 0$ for all A, B from $B_*^+(\mathcal{H})$, such that $\langle A|B\rangle = \mathrm{Tr}(AB) = 0$. Since $\mathbb{K}$ is linear it is enough to take $A = |\psi\rangle\langle\psi|$ on $B = |\varphi\rangle\langle\varphi|$. Now we obtain $\langle\, |\psi\rangle\langle\psi|, |\varphi\rangle\langle\varphi|\,\rangle = 0$ iff $\langle\varphi|\psi\rangle\langle\psi|\varphi\rangle = 0$, that is if $|\psi\rangle \perp |\varphi\rangle$, so we obtain

$$\langle\, |\psi\rangle\langle\psi|, \mathbb{K}(|\varphi\rangle\langle\varphi|)\,\rangle = \langle\varphi|\left(\mathbb{K}|\psi\rangle\langle\psi|\right)|\varphi\rangle \geq 0 \tag{48}$$

for all $|\psi\rangle \perp |\varphi\rangle$. Using the results from [1] and [2] we obtain the condition (47).

As it was shown in [32], some necessary and sufficient conditions for a generator of a quantum dynamical semi-group can also be formulated using the concept of dissipative operators in the sense of Lumer and Phillips [33]. Namely, if a sequence $\{\mathcal{P}_1, \mathcal{P}_2, \ldots\} \equiv \pi$ of projection operators on closed subspaces of the Hilbert space $\mathcal{H}$ constitutes a discrete resolution of identity, that is, if

$$\mathcal{P}_i \mathcal{P}_j = \delta_{ij} \mathcal{P}_j\,, \quad \text{and} \quad \sum \mathcal{P}_i = \mathbb{I}\,, \tag{49}$$

then one can describe the properties of the operator $\mathbb{K}$ in the following: a linear operator $\mathbb{K}$ generates a dynamical semigroup iff for every discrete reduction of identity we have

$$a_{ij}(\pi) \geq 0 \quad \text{for} \quad i \neq j, \qquad a_{ii}(\pi) \leq 0\,, \tag{50}$$

and

$$\sum_i a_{ij}(\pi) = 0 \tag{51}$$

where

$$a_{ij}(\pi) = \mathrm{Tr}(\mathcal{P}_i(\mathbb{K}\mathcal{P}_j)) \tag{52}$$

for $i, j = 1, 2, \ldots$. The conditions (50) are quantum analogs of Kolmogorov conditions (cf. [34]) for discrete Markov processes.

6. Summary

In Sections 4 and 5 it was shown that one can analyze the properties of superoperators which are important in modeling of open quantum systems using the natural isomorphisms defined by relation (28). In some situations this approach gives more effective results than other methods. On the other hand, other approaches (for example, the method proposed by Kossakowski, which is based on dissipative operators) give us a beautiful similarity to classical results of Kolmogorov.

Acknowledgement

This research has been supported by grant No. DEC-2011/02/A/ST1/00208 of National Science Center of Poland.

Author details

Andrzej Jamiołkowski

Institute of Physics, Nicolaus Copernicus University, Toruń, Poland

References

[1] A. Jamiołkowski, Rep. Math. Phys. 3, 275 (1972); ibidem 5, 415 (1975);

[2] A. Jamiołkowski, *On maps in the set of mixed states and quantum information thermodynamics*, Thesis (in Polish), 1972.

[3] I. Bengtsson, K. Życzkowski, *Geometry of Quantum States*, Cambridge Univ. Press, 2008.

[4] V. Vedral, *Introduction to Quantum Information Science*, Oxford Univ. Press, 2006.

[5] A. Holevo, *Statistical Structure of Quantum Theory*, Springer, 2001.

[6] D. Bruss, G. Leuchs (Eds.), *Lectures on Quantum Information*, Wiley-VCH, 2007.

[7] Z. Meglicki, *Quantum Computing without Magic*, MIT Press, 2008.

[8] F. Benatti, *Dynamics, Information and Complexity in Quantum Systems*, Springer, 2009.

[9] K. Kraus, Ann. Phys. 64, 119 (1971).

[10] A. Kossakowski, Rep. Math. Phys. 3, 243 (1972).

[11] O. Perron, Math. Ann. 64, 248 (1907).

[12] G. Frobenius, S. B. Preuss. Akad. Wiss. (Berlin), 256 (1912).

[13] J. S. Vandergraft, SIAM J. App. Math. 16, 1208 (1968).

[14] M. A. Nielsen, I. Chuang, *Quantum Computation and Quantum Information*, Cambridge Univ. Press., 2010.

[15] G.P. Barker, Linear Alg. Appl. 39, 263 (1981).

[16] R.S. Varga, *Matrix Iterative Analysis*, Springer, 2000.

[17] A. Berman, R.J. Plemmons, *Nonnegative Matrices in the Mathematical Sciences*, Academic Press, 1979.

[18] H. Wolkowicz et al., Eds., *Handbook of Semidefinite Programming*, Kluwer, 2003.

[19] A. Jamiołkowski, Open Sys. Infor. Dyn. 11 385 (2004).

[20] W. F Stinespring, Proc. AMS 6, 211 (1955).

[21] M. D. Choi, Linear Alg. Appl. 10, 285 (1975); ibidem 12, 95 (1975).

[22] R. M. Timoney, Bull. London Math. Soc. 32, 229 (2000).

[23] D. R. Farenick, Proc. AMS 124, 3381 (1996).

[24] D. Shemesh, Linear Alg. Appl. 62, 11 (1984).

[25] Yu. A. Alpin, Kh. D. Ikramov, J. of Math. Sciences 114, 1757 (2003).

[26] S. A. Amitsur, J. Levitzki, Proc. AMS 1, 449 (1950).

[27] V. Drensky, E. Formanek, *Polynomial Identity Rings*, Birkhäuser, 2004.

[28] Y. Alpin et al., Linear Alg. App. 312, 115 (2000).

[29] A. Jamiołkowski, Rep. Math. Phys. 46, 469 (2000).

[30] L. Elsner, Lin. Alg. Appl. 8, 249 (1974).

[31] A. Jamiołkowski, in preparation.

[32] A. Kossakowski, Bull. Acad. Polon. XX, 1021 (1972).

[33] G. Lumer, R. S. Phillips, Pacific J. Math. 2, 697 (1961).

[34] A. N. Kolmogorov, Math. Annal. 104, 415 (1931).

Chapter 2

Computational Complexity in the Analysis of Quantum Operations

Miłosz Michalski

Additional information is available at the end of the chapter

1. Introduction

Mathematical theory of infinite dimensional Hilbert spaces and the theory of operator algebras acting in such spaces (or C^* algebras in a more abstract approach) provide a standard setting for the formulation of modern quantum mechanics. On the other hand, experimental and theoretical progress achieved in the field of quantum information theory in the last two decades has indicated the practical and technological importance of low-dimensional quantum systems, where only a few basic modes play a significant role. Such modes can often be effectively decoupled from the rest of the system and controlled separately, providing physical realizations of qubits, qutrits and other basic information carriers. Regardless of concrete physical realization, be it photon polarization, electron or nuclear spin, charge in Josephson junctions to name just a few, the mathematical description of such systems requires only finite-dimensional Hilbert space language and finite-dimensional matrix algebras. Such structures are in principle computationally manageable in sharp contrast to the infinite dimensional ones.

It has to be pointed out, however, that there is a lot of misconception concerning the above mentioned "manageability" notion in today's quantum information literature. For instance, one of the most fundamental errors appearing in innumerable papers is to indiscriminately resort to the spectral resolution technique for hermitian matrices. Such an operation cannot be considered computationally effective if the size of the matrix exceeds 4: then it unavoidably involves solving an algebraic equation of degree 5 or more. Such task can be achieved only by an approximate numerical process, and therefore any emerging questions can be answered only up to numerical precision. The latter can be critical, for example, in checking whether a hermitian matrix has a negative eigenvalue.

Fortunately, in many situations similar to the one just described there are other alternative ways to obtain a *precise* answer, avoiding the approximate numerical computations. This is achieved by limiting oneself to the so-called finite *rational* computational procedures, involving only finitely many arithmetic operations on initial data so that the data as well as all intermediate and final computational results belong to the same number field. In particular, the use of transcendental functions is thus excluded.

The present chapter will be devoted to a review of a few such procedures, important for applications in quantum information theory. We will concentrate on the questions concerning not only the effectiveness of such procedures, but also on more detailed computational complexity issues. To describe better the subject of our considerations and to fix the terminology, let us consider the already invoked example of checking whether a given selfadjoint matrix has a negative eigenvalue, which in particular is a crucial ingredient in entanglement detection procedures. Note that the problem is posed so that the precise knowledge of the eigenvalue is not essential, it is its sign that matters.

Let A be a hermitian matrix in question and $\mathcal{H}$ be the respective Hilbert space. One can formulate the negative eigenvalue problem in an equivalent form by asking whether A is or is not positive semidefinite. As it is well known, positive semidefiniteness can be characterized by several equivalent criteria, each of them being an example of a different effectiveness or complexity issue. The list of relevant criteria is the following.

1. For each normalized vector $|\psi\rangle \in \mathcal{H}$ one has $\langle\psi\,|\,A\psi\rangle \geq 0$. The test based on this criterion is *ineffective* as it involves infinite number of conditions to verify, one for each $|\psi\rangle$.
2. All eigenvalues of A are nonnegative. As we have argued above, such test cannot be considered an effective one either. In general, the correctness of the answer hinges upon the numerical precision being used. We can call such tests *asymptotically effective*, meaning that increased numerical accuracy can yield the definite yes/no answer, but no a priori fixed precision is sufficient for the correctness of the whole class of such tests.
3. All principal minors of A are nonnegative. This is certainly an *effective* criterion as it involves the evaluation of finitely many subdeterminants of A. The computation of a determinant itself is a finite rational procedure. Note however, that direct application of the present criterion requires the evaluation of nearly 2^n minors, n being the size of A. Although finite, this number grows very rapidly with n, making the test *inefficient*. In practical terms, it may easily take years to complete such a test on the fastest computers, even for A of moderate size. For example, if A results from an application of some entanglement test to a mixed state of a system composed of merely 6 qubits, then $n = 64$ and hence the number of minors to compute is about $2^{64} \approx 10^{19}$. Assuming that our computing device can evaluate 10^6 minors per second on average, the time required to complete such a test would be of the order of 10^5 years. We characterize the computational complexity of such procedures by saying that they are *nonpolynomial* in n. Problems for which only nonpolynomial solution methods are available are termed *intractable*.
4. While the test of positive definiteness (Sylvester's criterion) is much simpler, for it involves only n *leading* principal minors of A, it has no counterpart for positive semidefinite matrices. However, one can easily check that the following recursive procedure based on Gaussian elimination can be used in this case. By A_{11} we denote here the submatrix of $A = [a_{ij}]$ obtained by the deletion of its 1st row and 1st column.
   (a) If $a_{11} < 0$ then A is *not* positive semidefinite.

(b) If $a_{11} = 0$ then A is *not* positive semidefinite unless its entire 1st column is null and A_{11} is positive semidefinite.

(c) If $a_{11} > 0$ then we first perform row-elimination of the entire 1st column of A. Then A is positive semidefinite iff the resulting A_{11} is such.

This is again a finite rational procedure. The largest computational effort in completing such a check is needed when there are no 0 entries in the first column of A and, likewise, no zeros are produced in A_{11} by the elimination. Then the recursive check uses the variant (c) repeatedly, so that the total number of arithmetic operations performed is of the order of n^3. The complexity of the method is thus *polynomial* and its efficiency is much higher than that of criterion 3. If as before $n = 64$, the test will complete in less than 1 second, assuming the computer speed of 10^6 rational arithmetic operations per second.

Positive semidefiniteness is certainly a very simple issue, however the above example highlights a few characteristic aspects of computational complexity. Mathematical problems often admit many different solution methods which, similarly as in our example, may range from ineffective to very efficient ones. Effective procedures however can often prove useless in practice if the computational effort involved grows too fast with the size of input data. The complexity of problems themselves can be characterized relative to the most efficient solution methods known for them. In some cases theoretical complexity bounds can be derived for classes of problems.

In the next section we provide a brief review of fundamental notions of computational complexity theory.

2. Basic notions of computational complexity theory

In theoretical computer science, algorithms are classified according to their *time* or *space complexity*. Time complexity gives an estimate of how does the number of elementary steps in the algorithm scale with the size of input data defining an instance of the problem. Space complexity refers to the scaling of the amount of workspace or extra memory (in one convention the memory storing input data is not counted) needed in the course of computation. The complexity of *problems* is related to their inherent difficulty and is a theoretical estimate of the computational cost indispensable for their solution. Often only some lower or upper complexity bounds are known for classes of problems. The complexity theory uses the formalism of abstract Turing machines to ensure the universality of conclusions.

It is not our goal to review the complexity theory in its general abstract formulation here, but rather to provide necessary intuitions for an unacquainted reader. Those familiar with computational complexity may well skip the current section.

The scaling of solution time or workspace with problem size is expressed using the "big O" notation.

DEFINITION 1. *For two functions $f, g\colon \mathbb{N} \to \mathbb{R}$ one writes $f(n) = O(g(n))$ for $n \to \infty$ if and only if*

$$\exists\, M \in \mathbb{R} \text{ and } n_0 \in \mathbb{N} \quad \text{such that} \quad |f(n)| \leq M|g(n)| \quad \text{for } n > n_0 .$$

For example, standard square matrix multiplication requires $O(n^3)$ arithmetic operations, n being the matrix size. Since no extra memory beyond that for data storage is needed for performing the multiplication, the space complexity here is $O(n^2)$. The Fast Fourier Transform performs $O(n \log n)$ arithmetic operations on an n element data vector. Evaluation of a determinant directly from its definition would involve the summation of $n!$ terms, however more efficient method using Gauss elimination reduces the effort to $O(n^3)$ arithmetic operations. Evaluation of a permanent on the other hand appears more complex (except for the case of computations over $\mathbb{Z}_2$, where $-1 \equiv 1(\text{mod}2)$ and hence $\det A = \text{per}\, A$): the best methods known so far [8, 23] have the complexity of $O(n2^n)$.

One of the objectives of the theory is to identify *complexity classes*, consisting of problems which can be solved by using only limited type of computational resources, which are abstractly characterized by restricted classes of Turing machines, most notably the classes P and NP. The class P consists of problems which can be solved by a deterministic Turing machine executing a number of steps bounded by a polynomial in the input data size. The class NP on the other hand consists of problems solvable in polynomial time by a *nondeterministic* Turing machine. As the latter can be simulated by a deterministic machine in exponential time, NP is often conventionally (yet not quite correctly) identified with the class of problems solved by exponential (nonpolynomial) time deterministic algorithms. Strictly speaking however, the essential feature of NP problems is that given a random candidate for a solution it takes no more than *polynomial* number of steps to verify its correctness or to reject it. Exponential time deterministic algorithms in NP can be thought of as performing an extensive "blind" search in the space of potential solutions (which is the actual source of nonpolynomial complexity) checking each of them at low (i.e. polynomial time) cost. In contrast, problems in P admit "clever" constructive solution methods.

In practical terms, problems of type P can be solved relatively fast regardless of their size, while for the NP type ones solution times become impractically long even for moderate size of input data, c.f. our discussion of positive semidefiniteness verification in the previous section. The distinction between efficient and inefficient methods is often used as a synonym for that between P and NP classes.

Obviously $\text{P} \subset \text{NP}$, but it is a famous open question (although today hardly believed to hold true) whether $\text{P} = \text{NP}$. The quest for an answer to the latter has led to the definition of various special complexity classes, in particular the class of NP-complete problems, NP-C. We say that a problem π can be polynomially transformed to another problem π', in written $\pi \propto \pi'$, if the solution of π for input data of size n can be obtained by means of the execution of an algorithm for π' at most a polynomial in n number of times on new data translated from the original input with at most polynomial effort. So if $\pi' \in \text{P}$ (resp. in NP), then any π such that $\pi \propto \pi'$ is necessarily also in P (resp. NP).

Definition 2. *A problem π is* NP-*complete iff*

(i) $\pi \in \text{NP}$;

(ii) $\forall \sigma \in \text{NP} \quad \sigma \propto \pi$.

If π satisfies only condition (ii), it is said to be NP-*hard.*

It may appear that NP-C can well be empty, but it is not so as shown by Cook in [7]. The first NP-C problem identified by Cook was the satisfiability of Boolean functions: given a Boolean function F in variables $x_1, \ldots, x_n$ doest there exist a truth/false assignment to all x_i making the value of F true? Cook's proof gives a method of how to cast, at polynomial cost, an arbitrary nondeterministic Turing machine into the one computing Boolean functions.

Knowing at least one NP-C problem it becomes easier to identify other ones: if $\pi \in$ NP is such that $\sigma \propto \pi$ for some $\sigma \in$ NP-C, then $\pi \in$ NP-C. The list of known NP-complete problems exceeds now 3000 items. By definition, providing a polynomial time solution to any single NP-C problem would automatically prove that P = NP. Because of this, NP-complete problems are considered the hardest among NP ones. In other words, it is generally believed that the search for exact polynomial time solution methods of NP-C problems is a waste of time. On the other hand, there are numerous problems of practical interest for which neither a proof of NP-completeness nor an efficient polynomial time solution method are known. The most notable example is the problem of finding factors of large integers.

It is interesting that a large class of problems in matrix theory which possess an efficient solution can be reduced to an evaluation of a small number of determinants or, equivalently, can be expressed, as above, in terms of Gaussian elimination or — still more elementary reduction — by a series of matrix multiplications. This point of view motivates the interest in the design of fast matrix multiplication algorithms. Perhaps the best known schema of this kind is due to V. Strassen (1969) and its complexity is $O(n^{2.81})$, while more recent method of Coppersmith and Winograd (1987) improves the efficiency to $O(n^{2.367})$, the theoretical lower bound being $O(n^2)$.

An example of NP-complete matrix algebra problem is the following [5]: given an $n \times m$ matrix A over $\mathbb{Z}$ with $n \leq m$, decide whether there exist a vanishing $n \times n$ subdeterminant of A. The evaluation of a permanent is NP-hard, for it is most likely not in NP class. Again, the existence of a polynomial algorithm for the computation of per A would infer the equality P = NP. Many complicated counting problems in combinatorics and graph theory can be reduced to an evaluation of a permanent. Actually, permanent evaluation is #P-complete, meaning that all counting functions which can be defined in terms of NP problems can be polynomially reduced to it, [25].

Another important complexity category, from a physicist's point of view, is the so-called BPP class (bounded error probabilistic polynomial time) consisting of decision problems solvable in polynomial time by a *probabilistic* Turing machine, with the probability of producing wrong answer bounded from above by a constant $0 \leq p < 1/2$. Less formally, this class corresponds to Monte Carlo algorithms likely to yield correct answers and running in polynomial time. Such conditions guarantee that in practice one can perform a relatively short series of independent runs of the method to learn the correct answer with very high probability. By Chernoff bound, the probability that incorrect answer appears in a series of runs most of the time decays exponentially with the series length. If instead of probabilistic one uses *quantum* Turing machines, the resulting class is called BQP (bounded error quantum polynomial time). It is shown that BPP $\subset$ BQP, but little is known so far about the relation of either of the classes to NP.

Finally, PSPACE is a class of problems solvable by deterministic Turing machines using at most polynomial in the data size amount of workspace. It is proved that adding nondeterminism does not alter this class, namely PSPACE = NPSPACE. NP is thus clearly

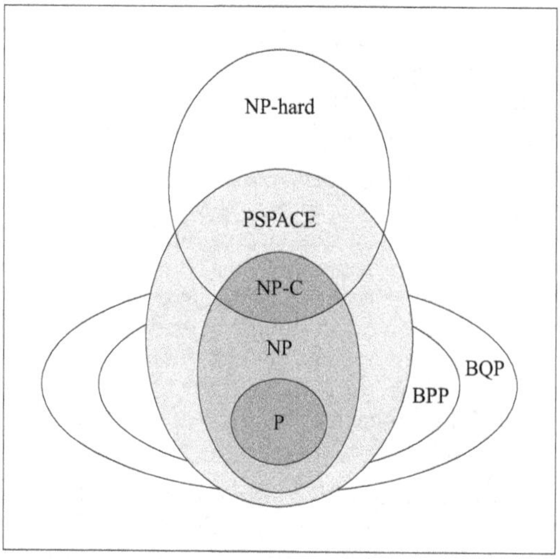

Figure 1. Hypothetical relations among complexity classes.

contained in PSPACE since using workspace of nonpolynomial size would automatically require nonpolynomial time. Fig. 1 summarizes what has been said above about the complexity classes.

Last but not least, there are problems which are provably undecidable, meaning that no finite algorithm can ever resolve them. Among such tasks there is the fascinating tiling problem [26].

Let us mention also that to date no general effective criteria are known for one of the most fundamental decision problems in quantum information, namely the determination whether a given mixed state of a bipartite system is entangled or not. All known exact methods, apart from those for low-dimensional systems, namely for $n = 4 = 2 \times 2$ and $n = 6 = 2 \times 3$, involve infinite number of computational tests (local actions of positive maps or, equivalently, evaluation of expectations of entanglement witnesses). Moreover, no effective method is in sight despite the two decades of intensive research efforts worldwide.

3. Some computational problems of quantum information theory

Quantum information (QI) theory regards quantum states as information carriers and quantum evolution of states as acts of information processing. As we have already mentioned in the Introduction, QI research focuses on low-dimensional quantum systems, qubits, qutrits and likewise, which appear to be most interesting from the point of view of potential future large-scale technological applications. Such low dimensional structures can be combined into multipartite quantum systems, realizing quantum registers and memories. Namely, given a low-dimensional Hilbert space, e.g. $\mathcal{H}_2 \simeq \mathbb{C}^2$ for a qubit, the space of the compound multipartite system is then

$$\mathcal{H} = \mathcal{H}_2 \otimes \cdots \otimes \mathcal{H}_2 = \mathcal{H}_2^{\otimes n} \simeq \mathbb{C}^{2^n}.$$

Genuinely quantum properties of such systems, most importantly the *entanglement* of their states, are proved to underlie the extraordinary efficiency of quantum information processing, surpassing that of the classical one. In what follows we shall silently assume finite-dimensionality of all quantum systems in question.

Let us recall that *pure states* of a quantum systems are represented by vectors in the respective Hilbert space, $|\psi\rangle \in \mathcal{H}$, while *observables*, i.e. measurable physical quantities, correspond to selfadjoint operators acting on $\mathcal{H}$, i.e. $A \in \mathcal{B}(\mathcal{H})$ such that $A = A^*$. In the finite-dimensional setting they can be identified with Hermitian matrices in the matrix algebra $\mathcal{M}_n(\mathbb{C})$, $n = \dim \mathcal{H}$. In passing to *mixed states* one replaces pure states with the corresponding 1-dimensional projection operators, $|\psi\rangle\langle\psi| \in \mathcal{B}(\mathcal{H})$, and one defines the mixed states as statistical sums of mutually orthogonal projections, $\varrho = \sum p_i |\psi_i\rangle\langle\psi_i|$ with real positive p_i summing up to 1. So defined, mixed states are quantum counterparts of classical discrete probability distributions. Their representatives are called density matrices. It can be easily seen that density matrices form a convex subset $\Sigma = \Sigma(\mathcal{H})$ of $\mathcal{B}(\mathcal{H})$ characterized by positive semidefiniteness and normalization of trace[1]

$$\varrho \in \mathcal{B}(\mathcal{H}) \qquad \text{such that} \qquad \varrho \geq 0 \ \text{ and } \ \operatorname{Tr} \varrho = 1.$$

According to the postulates of quantum mechanics, dynamical evolution of quantum systems is described by the Schrödinger equation, which, when reformulated for mixed states, takes the form of von Neumann equation

$$\dot{\varrho} = -i[H, \varrho] = -i(H\varrho - \varrho H).$$

Here H denotes the Hamiltonian of the system in question and we have assumed the convention $\hbar = 1$. This equation is solved by

$$\varrho(t) = U(t)\varrho(0)U^*(t),$$

where the unitary propagator has the form $U(t) = e^{-iHt}$.

Often, when the continuous time dependence of the system state is not the main issue, one resorts to discretized dynamics, using e.g. the "time one" mapping, $\varrho' = U\varrho U^*$. It turns out that general *quantum operations*, providing an adequate mathematical description of complex

[1] More consistently, mixed states should be regarded as elements of the Hilbert-Schmidt *dual* of $\mathcal{B}(\mathcal{H})$, that is linear functionals on $\mathcal{B}(\mathcal{H})$ acting on observables of the system by expectation $\varrho(A) = \operatorname{Tr}(\varrho A)$. For finite-dimensional $\mathcal{H}$ both $\mathcal{B}$ and $\mathcal{B}^*$ are in fact identical with $\mathcal{M}_n(\mathbb{C})$, the algebra of complex $n \times n$ matrices.

multi-stage quantum processes, experiments or computations acting on system states have a more general form of an operator sum

$$\Phi(\varrho) = \sum K_i \varrho K_i^* . \tag{1}$$

These include, for instance, quantum measurements or transmission of states through noisy quantum channels. The above so-called *Kraus representation* is the most general form of a linear *completely positive map* $\Phi : \mathcal{B}(\mathcal{H}) \to \mathcal{B}(\mathcal{H})$. From the point of view of quantum theory we are interested in the restriction of Φ to the set of density matrices $\Sigma(\mathcal{H})$. Complete positivity of Φ ensures that it preserves positivity of states, while an extra assumption is needed to guarantee the preservation of trace, namely $\sum K_i^* K_i = I$, where I denotes the identity matrix. So, for such Φ we have $\Phi : \Sigma \to \Sigma$. In QI theory such maps represent general quantum communication channels and typical questions studied in this context concern e.g. the effect of Φ on the initial entanglement of the transmitted states, the impact of noise, decoherence, etc. Let us mention also that Kraus representation, though very useful, has the defect of not being unique for a given quantum map Φ.

It should be stressed that quantum operations in the above sense are as a rule nonunitary. Even in the simplest case of Φ represented by two unitary (up to scaling) Kraus terms, $\Phi(\varrho) = U\varrho U^* + V\varrho V^*$, the action of Φ is *not* unitary unless $U = V$ up to a constant factor. However, this is does not pose a contradiction with postulates of quantum mechanics. Let us sketch briefly a typical *open system* scenario leading to nonunitary dynamics.

Suppose that we realistically consider a quantum system not as isolated one, but as remaining in contact with an external bath, so that the underlying Hilbert space has the structure $\mathcal{H} = \mathcal{H}_S \otimes \mathcal{H}_B$, with $\mathcal{H}_S$ and $\mathcal{H}_B$ being respectively the system and the bath spaces. It is natural then to cast the overall Hamiltonian in the following form:

$$H = H_S \otimes I_B + I_S \otimes H_B + H_I ,$$

where H_S and H_B are the Hamiltonians describing the evolution of the system and bath alone, H_I represents the interaction between them and I_S, I_B are the respective identity operators. While the overall system dynamics is unitary

$$\varrho(t) = U(t)\varrho(0)U^*(t) , \qquad U(t) = e^{-iHt} ,$$

it is intractable in such an exact form due to typically huge number of degrees of freedom of the bath. It is then natural to pass to a statistical description of the system evolution by averaging the bath out, assuming in addition that initially the system and the bath are decoupled, that is

$$\varrho_S(t) = \mathrm{Tr}_B\Big(U(t)\,\varrho_S(0) \otimes \varrho_B(0)\,U^*(t)\Big) = \sum_\alpha A_\alpha(t)\varrho_S(0)A_\alpha^*(t) , \tag{2}$$

where the Kraus operators emerge as $A_\alpha = c_\alpha \langle \beta_i | U | \beta_j \rangle$ with α enumerating index pairs (i, j) and $|\beta_i\rangle$ being the bath basis states. This is clearly a nonunitary evolution unless all A_α are the same up to scalar factors — an unlikely event.

Nevertheless, there may exist a smaller subspace $\mathcal{H}_{DF}$ of $\mathcal{H}_S$ where the reduced dynamics (2) actually *is* unitary. This is equivalent to saying that there exists a basis of $\mathcal{H}_S$ in which all Kraus operators A_α have simultaneously the block form

$$A_\alpha = \left[\begin{array}{c|c} s_\alpha V & 0 \\ \hline 0 & \tilde{A}_\alpha \end{array}\right], \tag{3}$$

where V is unitary on $\mathcal{H}_{DF}$, s_α are scaling factors and $\tilde{A}_\alpha$ are arbitrary operators on $\mathcal{H}_{DF}^{\perp}$, the orthocomplement of $\mathcal{H}_{DF}$ in $\mathcal{H}_S$. Such a space is called *decoherence-free* as the coherent state evolution in this space is isolated from the destructive impact of the bath.

Similarly, one can derive conditions for the existence of a decoherence-free subspace in the framework of Markovian approximation of an open system dynamics, and they turn out to have a form consistent with (3) above. Let us recall that the following master equation in the Gorini-Kosakowski-Sudarshan form provides the most general description of a completely positive Markovian time evolution of a quantum system interacting with its environment [11, 20],

$$\dot{\varrho} = -i[H,\varrho] + \frac{1}{2}\sum_{ij} c_{ij}\Big([F_i, \varrho F_j^*] + [F_i\varrho, F_j^*]\Big), \tag{4}$$

where the sum collects all the terms responsible for nonunitary decohering dynamics. Thus H is the system Hamiltonian, the operators F_i are the so-called error fields and they represent the coupling of the system with its environment, while the hermitian structure matrix $[c_{ij}]$ carries other physically relevant information. Now, if $\mathcal{H}_{DF}$ is to be a decoherence-free subspace, then for any ϱ supported on it the second term in (4) must vanish identically, so that the resulting dynamics is purely unitary. If one assumes certain robustness, or *generic property* in the terminology of [18], of this subspace, meaning that the vanishing of the nonhamiltonian part is not the result of some fine-tuning among structure parameters c_{ij} but rather the effect of simultaneous vanishing of all individual terms, it can be seen that $\mathcal{H}_{DF}$ must be spanned by common eigenvectors of all error fields. In particular, $[F_i, F_j] = 0$ on $\mathcal{H}_{DF}$.

Let us now go back to general quantum operations represented by completely positive trace preserving maps in the form (1). As we have seen, the basic issue in the search for decoherence free subspaces is the identification of common eigenvectors of all Kraus operators K_i and maximal common invariant subspaces spanned by them. For reasons outlined in the introduction, it is impractical to approach this problem by means of direct evaluation of eigenvectors. As a rule, such computations are prone to numerical errors and hence the precise identification of common eigenvectors cannot be achieved this way. In section 5, we will describe an alternative constructive method based on simple linear algebra, the so-called Shemesh criterion, which allows one to identify common invariant subspaces of several operators.

We shall conclude this section by mentioning three more situations where the identification of common invariant subspaces plays a significant role.

1. Characterization of irreducibility of quantum operations, [14, 15]. Irreducible quantum operations (superoperators) appear as a natural generalization of the notion of positive semidefinite irreducible linear operators, treated in particular by Perron-Frobenius theory. The latter provides a very useful and simple characterization of the spectra of irreducible operators. It turns out that if a quantum operation Φ is given in terms of Kraus representation (1), then it is irreducible if and only if the operators K_i do not share a nontrivial invariant subspace. In other words, no decoherence-free subspace exists for an irreducible Φ.
2. Identification of sufficient algebras of observables, [12, 13]. To identify an unknown quantum state ϱ, an experimenter has to perform a number of measurements on the system in question, collecting data that can be used subsequently in the estimation of ϱ. Each of these measurements returns an expectation of the measured observable A_i in the state ϱ, that is the quantity $\mathrm{Tr}(A_i\varrho)$. A natural question that emerges is how to optimize such a data collection, namely how to choose observables A_i to obtain maximum information with the least experimental effort. Sufficiency of an algebra generated by a finite collection of observables $\mathcal{A} = \mathcal{A}(A_1, \ldots, A_p)$ means that the information acquired in the measurement process $\mathrm{Tr}(A_i\varrho)$, $i = 1, \ldots, p$, characterizes the state ϱ completely. One of the rationally verifiable conditions which can be used here is based on Burnside's theorem, which allows one to check whether a given set of observables generates the full matrix algebra $\mathcal{M}_n$ or not. This question can again be related with the existence of a common invariant subspace for the generators of $\mathcal{A}$.
3. Error correcting codes, [6, 17]. This is a more general case than that of the existence of a decoherence-free subspace. Here, one is interested in establishing the existence of a subspace $\mathcal{H}_{EC}$, the subscript EC for *error correcting*, of $\mathcal{H}_S$ on which the action of the channel Φ can be effectively inverted, namely, there exists a quantum operation Θ such that for states ϱ supported on $\mathcal{H}_{EC}$ one has $\Theta(\Phi(\varrho)) = \varrho$. The motivation behind such a demand is that the basis states of $\mathcal{H}_{EC}$ can be regarded as "code words" which can unambiguously be unscrambled after transmission through the generally corrupting channel Φ, and thus they can be used to safely encode portions of information to be sent through the channel. As shown in [17], the necessary and sufficient condition for the existence of an EC subspace for an operation Φ resulting from (2) can be phrased in the following simple algebraic form involving the Kraus operators A_α: there exists a basis of $\mathcal{H}_S$ such that for all α, β

$$A_\alpha^* A_\beta = \left[\begin{array}{c|c} r_{\alpha\beta} I & 0 \\ \hline 0 & \tilde{A}_\alpha^* \tilde{A}_\beta \end{array}\right],$$

where as before $\tilde{A}_\alpha, \tilde{A}_\beta$ are arbitrary operators on $\mathcal{H}_{EC}^\perp$ and $R = [r_{\alpha\beta}]$ is a scalar matrix. I in the upper left block is the identity on $\mathcal{H}_{EC}$. Note that the decoherence-free subspace is a special case of an EC space, since then from (3) it follows that the matrix R has a very special form $r_{\alpha\beta} = \bar{s}_\alpha s_\beta$ and therefore has rank 1.

4. Characteristic and minimal polynomials

As we have mentioned in the introduction, the precise determination of eigenvalues of a matrix by means of a finite rational computation is in general impossible. The same is true for eigenvectors. One can nevertheless rationally acquire exact knowledge about some spectral properties of a matrix, for instance by studying its characteristic and minimal polynomials. Numerous methods for obtaining the polynomials can be found in algebraic literature, and we are going to recall two of them here.

For an $n \times n$ complex matrix A let

$$\chi_A(\lambda) = \det(\lambda I - A) = \lambda^n + p_1\lambda^{n-1} + \cdots + p_{n-1}\lambda + p_n$$

be its characteristic polynomial. We will describe the method of undetermined coefficients — an efficient algorithm yielding the numbers p_i. The procedure begins with the evaluation of auxiliary constants

$$D_k := \chi_A(k) = \det(kI - A), \qquad k = 0, 1, \ldots, n-1.$$

Next the following system of linear equations in the unknowns $p_1, \ldots, p_n$ is formed

$$\begin{cases} & & & p_n = D_0 \\ 1^n & + & p_1 1^{n-1} & + \cdots + p_n = D_1 \\ 2^n & + & p_1 2^{n-1} & + \cdots + p_n = D_2 \\ \cdots & & & \cdots \quad \cdots \\ (n-1)^n & + & p_1(n-1)^{n-1} & + \cdots + p_n = D_{n-1} \end{cases}$$

or equivalently

$$\begin{bmatrix} 1^{n-1} & 1^{n-2} & \cdots & 1 \\ 2^{n-1} & 2^{n-2} & \cdots & 2 \\ \vdots & \vdots & \ddots & \vdots \\ (n-1)^{n-1} & (n-1)^{n-2} & \cdots & n-1 \end{bmatrix} \begin{bmatrix} p_1 \\ p_2 \\ \vdots \\ p_{n-1} \end{bmatrix} = \begin{bmatrix} D_1 - D_0 - 1^n \\ D_2 - D_0 - 2^n \\ \vdots \\ D_{n-1} - D_0 - (n-1)^n \end{bmatrix}.$$

Writing S_{n-1} for the matrix on the left hand side, the solution can be expressed in compact vector notation as $p = S_{n-1}^{-1}D$. Note that S_{n-1} is a constant matrix whose inverse can be computed and stored beforehand and used repeatedly for various input matrices A. The computational cost is thus limited to the determination of the vector D, and hence is bounded by $O(n^4)$. For comparison, direct expansion expressing the coefficients p_i by the sums of i-th order principal minors of A results in the computation scheme of complexity $O(2^n)$.

The minimal polynomial of a A is defined to be the least degree monic polynomial μ (i.e. with the leading coefficient 1) which annihilates A, $\mu(A) = 0$. Alternatively, it can be given in the form

$$\mu_A(\lambda) = (\lambda - \lambda_1)^{r_1} \cdots (\lambda - \lambda_k)^{r_k},$$

where λ_i are distinct eigenvalues of A and r_i denotes the order of the largest Jordan block for λ_i in the canonical representation of A. Clearly μ_A divides χ_A.

One obvious direct method consists in checking the sequence of matrices

$$I,\ A,\ A^2,\ \ldots,\ A^r$$

for linear independence, systematically for $r = 1, 2, \ldots$. The least r for which the sequence turns out to be linearly dependent is the degree of the minimal polynomial μ_A, and the respective vanishing linear combination

$$c_r I + c_{r-1} A + \cdots + c_1 A^{r-1} + c_0 A^r = 0$$

yields, after dividing by c_0, the coefficients of μ_A. This task can be realized by applying Gauss elimination to the $r \times n^2$ matrix whose rows are the reshaped matrices $I, A, A^2, \ldots$, i.e. row vectors obtained by arranging the elements of A^i lexicographicaly row after row. The complexity of such process is $O(n^4)$.

An equivalent method often used in practice is a variant of Krylov subspace algorithm, based on the following classical theorem.

THEOREM 1. *For a linear map $A : V \to V$ let $W_1, \ldots, W_k$ be subspaces of V such that*

i) $W_1 + \cdots + W_k = V$, *the sum not necessarily being direct,*

ii) each W_i is invariant for A,

iii) the restriction $A|_{W_i}$ has minimal polynomial m_i.

Then the minimal polynomial μ_A of A on V is the least common multiple of $m_1, \ldots, m_k$.

The algorithm has the following steps.

1. Pick nonzero $v \in V$ and iteratively compute its Krylov subspace relative to A,

$$W = \mathrm{Span}\{v, Av, \ldots, A^{d-1}v\}.$$

   That is, d is the smallest number such that the vectors $v, Av, \ldots, A^d v$ are linearly dependent, namely

$$A^d v = c_1 A^{d-1} v + \cdots c_{d-1} Av + c_d v.$$

   By construction, the subspace W is invariant for A. It is not difficult to justify that

$$m(\lambda) = \lambda^d - c_1 \lambda^{d-1} - \cdots - c_{d-1}\lambda - c_d$$

   is the minimal polynomial of the restriction $A|_W$.

2. Set $W_1 = W$ and $m_1 = m$. If $W_1 = V$ we are done, otherwise pick $v' \notin W_1$ and repeat step 1 to obtain W_2 and m_2 and so on. The construction terminates when $W_1 + W_2 + \cdots + W_k = V$.

3. Find μ_A as the least common multiple of $m_1, \ldots, m_k$. This can be done rationally by using Euclid's algorithm repeatedly to find first GCD of pairs of polynomials m_i.

Most of the computational effort resides here in the construction of Krylov subspaces. For each new vector $A^i v$ added to W linear dependence is checked by Gaussian elimination. Altogether no more than n such checks are performed so the complexity bound is $O(n^4)$.

Let us conclude this section by mentioning some exemplary problems in quantum physics, where knowledge of spectral and minimal polynomials plays a role. Firstly, it is the design of optimal setups for stroboscopic tomography of states [12, 13]. Namely, one has to find a minimal set of observables and design a stroboscopic measurement, i.e. one performed at preselected time instants when the measured observables are subdued to time evolution, the objective being to collect information sufficient for the complete reconstruction of a quantum state with least experimental effort. To this end, Krylov subspaces of the observables relative to the generator of the dynamics have to be constructed. The degree of the minimal polynomial of the dynamics generator is one of the essential parameters appearing in the design process.

Second set of examples is related to the construction of common invariant subspaces for families of operators, which finds application e.g. in the identification of decoherence-free subspaces in open quantum systems. This problem will be discussed in detail in the next section. It turns out that the construction of such common invariant subspaces can be simplified considerably if one of the operators has nondegenerate spectrum. The former property can be tested for an operator A by analyzing the GCD of its characteristic polynomial and its derivative: the eigenvalues are simple iff χ_A and χ'_A are relatively prime. To detect diagonalizability, one has to perform a similar test on the minimal polynomial of A. An alternative for the Euclidean GCD algorithm is the singularity test of the so-called associated Sylvester matrix [27].

5. Common invariant subspaces

The problem we are going to discuss now in its simplest version can be formulated as follows: given two square matrices $A, B \in \mathcal{M}_n$ decide whether they have an eigenvector in common. We are interested, of course, in finite *rational* procedures solving this problem. As it was indicated in the introduction, naive direct approach by literally finding the eigenspaces of A and B and comparing them is useless because of finite accuracy of numerics. We will be concerned with a more general formulation of the problem, namely we will ask whether two matrices share an invariant subspace of dimension k and how to find such subspace.

In what follows, we will discuss certain finite rational computational procedures detecting the existence of common invariant subspaces for pairs of operators. There are no known direct generalizations of such procedures to work for more than two operators at a time. However, if one can constructively obtain common invariant subspaces for all pairs of operators in the set $A_1, \ldots, A_p$, then taking their intersection one obtains a candidate for the global solution. It has to be verified though, because the resulting space need not be invariant for some (or any!) of the operators A_i. The computational complexity of such a construction will add a factor p^2 to that of the process performed for a single pair of operators. The intersection of p^2 subspaces of dimensions bounded by n can be constructed in time bounded by $p^2 n^3$.

5.1. Shemesh criterion and related methods

The basic tool in the detection of common invariant subspaces is the so-called Shemesh criterion [24]. We use here the standard notation $[A, B]$ for the commutator of matrices A and B.

THEOREM 2 (Shemesh 1984). *Matrices $A, B \in \mathcal{M}_n$ possess a common eigenvector if and only if the subspace*

$$\mathcal{N} = \bigcap_{k,l=1}^{n-1} \ker[A^k, B^l] \tag{5}$$

is of positive dimension. Moreover, $\mathcal{N}$ is invariant with respect to both A and B and restrictions of A and B to $\mathcal{N}$ commute. Every common invariant subspace of A and B (on which they commute) is contained in $\mathcal{N}$.

Let us remark that n above can be replaced by r and s — the degrees of minimal polynomials of A and B, respectively.

We shall analyze now the complexity of a direct method of checking Shemesh criterion and that of constructing $\mathcal{N}$ — the maximal common invariant subspace of A and B. Let us stress here that while the existence of a 1-dimensional common invariant subspace (corresponding to the common eigenvector of A and B) in $\mathcal{N}$ is guaranteed by the criterion, it *does not* answer any questions concerning k-dimensional common invariant subspaces, $2 \leq k < \dim \mathcal{N}$, not to mention the problem of constructing them by finite rational procedures. Such procedure can be nevertheless easily obtained for the space $\mathcal{N}$. Let us also indicate that no finite rational method should be expected to yield the common eigenvector in $\mathcal{N}$. If there were one, we would have a finite method to compute exactly the corresponding eigenvalues of A and B which is, in general, unfeasible.

To estimate the complexity of Shemesh's criterion, let us first note that computing the commutator $[A, B]$ has the same complexity as matrix multiplication[2], namely $O(n^3)$. The number of commutators to evaluate in (5) is at most $(n-1)^2$, so that the total amount of algebra is bounded here by $O(n^5)$. Finally, finding the intersection of kernels can be done just by means of solving the system of homogeneous linear equations in n variables given by the $n(n-1)^2 \times n$ matrix

$$\begin{bmatrix} [A, B] \\ [A, B^2] \\ \vdots \\ [A^{n-1}, B^{n-1}] \end{bmatrix}. \tag{6}$$

This is achieved by the Gaussian elimination algorithm again in $O(n^5)$ steps, hence the overall complexity of finding $\mathcal{N}$ is $O(n^5)$.

[2] Of course, one can always lower the exponent 3 to some extent by resorting to fast matrix multiplication schemes. This may be of practical importance when working with large matrices, here however we are mainly interested in establishing just *polynomial* complexity of our procedures.

An equivalent formulation of Shemesh condition ($\dim \mathcal{N} > 0$) is that

$$\det \sum_{k,l=1}^{n-1} [A^k, B^l]^* \cdot [A^k, B^l] = 0$$

but it does not simplify the computation as the sum above involves $(n-1)^2$ terms, each one computable with the arithmetic cost of $O(n^3)$ operations.

Let us turn to a more complicated problem of verifying the existence of a common invariant subspace of prescribed dimension $2 \leq k < n$. This is partly solved by applying the Shemesh criterion to exterior powers (wedge powers) of A and B. Recall that $A^{\wedge k}$ is the restriction of $A^{\otimes k}$ to the antisymmetric subspace of $\left(\mathbb{C}^n\right)^{\otimes k}$. More explicitly, $A^{\wedge k}$ is an $m \times m$ matrix with $m = \binom{n}{k}$, the elements of which are

$$\left(A^{\wedge k}\right)_{\alpha,\beta} = \det A[\alpha|\beta],$$

where α and β stand for multi-indices $\alpha = (i_1, i_2, \ldots, i_k)$, with $1 \leq i_1 < i_2 < \cdots < i_k \leq n$. $A[\alpha|\beta]$ is a $k \times k$ submatrix of A with rows and columns specified by α and respectively β. The space $\mathcal{N}_k$ corresponding to $\mathcal{N}\,(= \mathcal{N}_1)$ in (5) is now defined by analogy as

$$\mathcal{N}_k = \bigcap_{i,j=1}^{m-1} \ker\left[\left(A^{\wedge k}\right)^i, \left(B^{\wedge k}\right)^j\right]. \tag{7}$$

The trick of using exterior algebra takes advantage of a simple fact that if $\lambda_1, \ldots, \lambda_k$ are eigenvalues of A with (linearly independent) eigenvectors $v_1, \ldots, v_k$ then $\lambda_1 \lambda_2 \cdots \lambda_k$ is an eigenvalue of $A^{\wedge k}$ with eigenvector $v_1 \wedge \cdots \wedge v_k$. So if $v_1, \ldots, v_k$ span an invariant k-dimensional subspace of A and B then obviously $v_1 \wedge \cdots \wedge v_k$ is a common eigenvector of $A^{\wedge k}$ and $B^{\wedge k}$. The corresponding sufficient condition, however, turns our to be more complicated. Nontriviality of $\mathcal{N}_k$ guarantees the existence of an eigenvector shared by $A^{\wedge k}$ and $B^{\wedge k}$ but it is now an object in the exterior algebra of $\mathbb{C}^n$ and, in general, it need not be decomposable, i.e. of pure product form $v = v_1 \wedge \cdots \wedge v_k$. Consequently the reconstruction of a k-dimensional common invariant subspace of A and B from v may no longer be easy if at all possible. The source of this difficulty resides in the fact that the spectrum of $A^{\wedge k}$ or $B^{\wedge k}$ may be degenerate. This possibility has to be, therefore, excluded by an additional assumption. As we will see shortly, such an assumption can be further relaxed to another one postulating the nondegeneracy of eigenvalues of either A or B alone.

The generalized Shemesh criterion [9] takes the following form.

THEOREM 3 (Generalized Shemesh Criterion).

NECESSITY: *If A and B have a common invariant subspace of dimension $2 \leq k < n$, then $\mathcal{N}_k$ as defined in (7) has positive dimension (i.e. $A^{\wedge k}$ and $B^{\wedge k}$ share an eigenvector).*

SUFFICIENCY: *Suppose that $A^{\wedge k}$ has nondegenerate eigenvalues and* $\det B \neq 0$. *Then if $\mathcal{N}_k \neq \{0\}$, there exists a common k-dimensional invariant subspace of A and B.*

In order to show how one can simplify the extra conditions in the sufficiency part of the above theorem, let us note that for an arbitrary matrix C the spectral shift transformation $C \mapsto C_t = C - tI$ does not alter its invariant subspaces. The following two facts proved in [9] allow one to preprocess, if necessary, the initial matrices A and B so that the extra requirements are fulfilled, at the same time leaving their invariant subspaces intact.

FACT 1. For any singular complex matrix B, a shift $t \in \mathbb{N}$ can be computed by a finite rational procedure so that $\det(B - tI) \neq 0$.

The procedure is very simple: it computes $\det(B - tI)$ for $t = 1, 2, \ldots$ until a nonzero value is found. Since the characteristic polynomial of B has no more than n distinct roots, the computation must terminate in no more than $n - 1$ steps.

FACT 2. If all eigenvalues of A are nondegenerate and $2 \leq k < n$, then a shift $t \in \mathbb{N}$ can be computed by a finite rational procedure so that the matrix $(A - tI)^{\wedge k}$ also has only simple eigenvalues.

See [9] for the proof of Fact 2. Its essence is that one can probe subsequent values of the shift parameter $t = 0, 1, \ldots$ until nondegeneracy of eigenvalues occurs, which is shown to happen after no more than $\frac{1}{2}kn^{2k}$ of such tests.

We are equipped now to describe the complete algorithm determining the existence of k-dimensional invariant subspace common to A and B. Let ϕ_A denote the characteristic polynomial of A.

1. Check whether A has distinct eigenvalues by computing the resultant of ϕ_A and ϕ'_A (as we have mentioned in Section 3, this can be done conveniently by expressing it as the determinant of the Sylvester matrix [27] of ϕ_A and ϕ'_A) and checking whether it is nonzero. If the test fails for A, try the same for B and switch A and B if B has simple eigenvalues. If both tests fail, the generalized Shemesh criterion cannot be used.
2. If B is singular, apply the spectral shift t as in Fact 1. Replace B with $B - tI$.
3. Compute the matrix $A^{\wedge k}$ and check whether it has nondegenerate eigenvalues (see step 1). If so, go to step 4, otherwise apply the spectral shift to A as described in Fact 2 and repeat step 3.
4. Compute $B^{\wedge k}$ and $\mathcal{N}_k$ as in (7). If $\mathcal{N}_k$ has positive dimension, then A and B have common k-dimensional invariant subspace.

It should be stressed again that Shemesh criterion yields a "yes/no" answer about the *extistence* of a common eigenvector (or, respectively, of k-dimensional common invariant subspace), but does not help in *constructing* them.

The complexity of the above algorithm is determined by n and k. The most time-consuming operations are those performed on the exterior powers of A and B because of their size $m = \binom{n}{k}$, which grows roughly like n^k for $k = 2, \ldots, \left[\frac{n}{2}\right]$. To obtain $A^{\wedge k}$, one has to evaluate m^2 minors of A of size $k \times k$, hence the computational cost is bounded by $O(k^3\, n^{2k})$. Checking for nondegeneracy of eigenvalues of $A \in \mathcal{M}_n$ costs as much as the evaluation of ϕ_A, which can be done in $O(n^3)$ steps, plus the cost of computing the $(2n - 1) \times (2n - 1)$ determinant of the respective Sylvester matrix, so its overall complexity is $O(n^3)$. Step 3 of the algorithm probing possible shift parameters performs no more than $O(km^2)$ of nondegeneracy tests, each at the expense of $O(m^3)$ arithmetic operations. Therefore the complexity of step 3 evaluates to $O(kn^{5k})$. Finally, the complexity of constructing $\mathcal{N}_k$ by (7) is, as shown before, $O(m^5)$ or in terms of n and k $O(n^{5k})$.

The estimation above shows that even for small values of k, although of polynomial time complexity, the method is not very practical. Already for $k = 2$, the computational effort is of the order $O(n^{10})$ in the worst case.

Let us mention one more recent result [16] which shows that the nondegeneracy condition can in fact simplify the original Shemesh criterion, slightly reducing its computational complexity.

THEOREM 4 (Jamiołkowski, 2012). *Let A have only simple eigenvalues. Then the formula (5) for the space $\mathcal{N}$ in the original Shemesh criterion can be simplified to*

$$\mathcal{N} = \bigcap_{k=1}^{n-1} \ker\left[A^k, B\right] \tag{8}$$

which reduces the complexity of its construction to $O(n^4)$.

Indeed, the number of commutators to evaluate in (8) is now at most $n-1$, $O(n^3)$ arithmetic operations each, and the system of homogeneous equations defining $\mathcal{N}$ is of the size $n(n-1) \times n$, so the complexity of solving it is also $O(n^4)$.

As the sufficiency part of the generalized Shemesh condition requires the nondegeneracy of the spectrum of $A^{\wedge k}$, so the formula (7) automatically simplifies analogously to

$$\mathcal{N}_k = \bigcap_{i=1}^{m-1} \ker\left[\left(A^{\wedge k}\right)^i, B^{\wedge k}\right]. \tag{9}$$

Hence the complexity of finding $\mathcal{N}_k$ reduces to $O(m^4)$, that is $O(n^{4k})$.

Let us note, however, that somewhat weaker assumption of diagonalizability of A does not, in general, lead to a simplification of the Shemesh formula by limiting the number of commutators that have to be computed. This is illustrated by the following simple example. Let $\{e_1, \ldots, e_4\}$ be a basis in which A and B have the following form:

$$A = \begin{bmatrix} 1 & 1 & & \\ 1 & 2 & & \\ & & 3 & \\ & & & 3 \end{bmatrix}, \qquad B = \begin{bmatrix} 2 & 1 & & \\ & 2 & 1 & \\ & & 2 & 1 \\ & & & 2 \end{bmatrix},$$

where we have suppressed all zero entries. Note that A is diagonalizable with twofold degenerate eigenvalue 3. Its minimal polynomial has degree 3. Hence

$$\ker[A,B] \cap \ker[A^2,B] \;=\; \ker[A,B] \cap \ker[A^2,B] \cap \ker[A^3,B] \;=\; \mathrm{Span}\{e_4\}$$

but

$$\ker[A,B] \cap \ker[A^2,B] \cap \ker[A,B^2] \;=\; \{0\}\,.$$

In the next subsection we will explore an alternative approach based on the so-called polynomial identities for matrix algebras.

5.2. Algebraic approach — polynomial identities

In algebra, polynomial identities are used to characterize various algebraic structures. We will limit the exposition to a necessary minimum so as to make the present text self-contained, focusing on applications to common invariant subspace problems.

Definition 3. *An algebra $\mathcal{A}$ is said to be a polynomial identity algebra (a PI-algebra for short) if there exists a polynomial $P(x_1, x_2, \ldots, x_k)$ over the ring of integers in noncommuting variables x_i such that $P(A_1, A_2, \ldots, A_k) = 0$ for all k-tuples of the elements A_i of $\mathcal{A}$.*

For example, a commutative algebra $\mathcal{A}$ is a PI-algebra with the polynomial $Q_2(x_1, x_2) = x_1x_2 - x_2x_1$. It turns out that special role is played by the so-called standard polynomials which are natural generalizations of Q_2,

$$Q_n(x_1, \ldots, x_n) \;=\; \sum_{\sigma \in S_n} \mathrm{sign}(\sigma) x_{\sigma(1)} \cdots x_{\sigma(n)}\,, \tag{10}$$

where the summation extends over the symmetric group S_n. Their importance is exemplified by the Amitsur-Levitzki theorem on matrix algebras $\mathcal{M}_n$.

Theorem 5 (Amitsur-Levitzki 1950). *The full algebra $\mathcal{M}_n(\mathbb{C})$ satisfies the standard polynomial identity of degree $2n$,*

$$Q_{2n}(A_1, \ldots, A_{2n}) \;\equiv\; 0 \qquad \forall\, A_1, \ldots, A_{2n} \in \mathcal{M}_n\,,$$

but it does not satisfy any polynomial identity of smaller degree.

In order to make a connection with the problem of common invariant subspaces, let us first observe that if two matrices A and B share such a subspace W, then W is also invariant for the entire algebra $\mathcal{A}(A,B) \subset \mathcal{M}_n$ generated by A and B. In what follows we shall denote this algebra by $\mathcal{A}$ for simplicity. So according to the Shemesh criterion (5), $\mathcal{A}$ restricted to $\mathcal{N}_1$ satisfies the standard polynomial identity $Q_2 \equiv 0$, that is

$$(C_1C_2 - C_2C_1)v = 0\,, \qquad \forall\, C_1, C_2 \in \mathcal{A}, \quad \forall\, v \in \mathcal{N}_1.$$

Following [2] let us define the family of subspaces

$$\mathcal{N}_k = \bigcap \ker \left[Q_{2k}(C_1, \ldots, C_{2k}) C_{2k+1} \right] , \tag{11}$$

where the intersection extends over all $(2k+1)$-tuples of elements $C_i \in \mathcal{A}$. It turns out that $\mathcal{A}$ restricted to $\mathcal{N}_k$ analogously obeys the identity $Q_{2k} \equiv 0$. Of course, this is an interesting property provided that $\mathcal{N}_k$ is not just the zero space.

THEOREM 6. *If $\mathcal{N}_k$ of (11) is nontrivial, then it is an invariant subspace for $\mathcal{A}$ and this algebra restricted to $\mathcal{N}_k$ satisfies the standard polynomial identity $Q_{2k} \equiv 0$, that is*

$$Q_{2k}(C_1, \ldots, C_{2k}) v = 0 , \qquad \forall\, C_1, \ldots, C_k \in \mathcal{A}, \quad \forall\, v \in \mathcal{N}_k.$$

Any other invariant subspace of $\mathcal{A}$ on which this algebra satisfies the identity $Q_{2k} \equiv 0$ is contained in $\mathcal{N}_k$.

The proof can be found e.g. in [2]. The usefulness of this theorem can be appreciated by noting that for subsequent values of k we obtain a filtration

$$\{0\} = \mathcal{N}_0 \subset \mathcal{N}_1 \subset \cdots \subset \mathcal{N}_n = \mathbb{C}^n ,$$

which can yield partial answers to questions concerning invariant subspaces of specific dimension. We stress here that each of $\mathcal{N}_k$ can be constructed by a finite rational procedure. Namely, because of linearity of Q_{2k} with respect to each individual variable, to find $\mathcal{N}_k$ it suffices to make each C_i in the intersection (11) run independently through the elements of a fixed basis of $\mathcal{A}$.

The basis itself can be found by the following general procedure [1]. Consider finite products of A and B, e.g. AB^2AB (called words over $\{A, B\}$) in lexicographic order:

$$I,\ A,\ B,\ A^2,\ AB,\ BA,\ B^2,\ A^3,\ A^2B,\ \ldots$$

Words of a fixed length k form the k-th layer in this sequence. I alone forms here the zeroth layer. Let $\mathcal{A}_k$ be the subspace of $\mathcal{M}_n$ spanned collectively by the layers $0 \leq j \leq k$. Obviously,

$$\mathcal{A}_0 \subset \mathcal{A}_1 \subset \cdots \subset \mathcal{A}_p = \mathcal{A}_{p+1}$$

for some p, the symbol $\subset$ denoting here the proper inclusion. Then $\mathcal{A}_p = \mathcal{A}(A, B)$ and the first $p+1$ layers form the spanning set for $\mathcal{A}$.

To discuss the complexity of this procedure, note first that an obvious rough bound for p is $p \leq n^2 - 1$, while there are various better estimates known in literature, see e.g. [10, 21, 22], especially when some knowledge about A and B is available. In particular, if A and B commute, then $p < n$, while the best general bound so far is that due to Pappacena [21],

$$p \;\leq\; n\sqrt{\frac{2n^2}{n-1} + \frac{1}{4}} + \frac{n}{2} - 2 \;\sim\; O(n^{3/2}) .$$

However, bad news is that the k-th layer contains 2^k words, so to construct $\mathcal{A}_k$ one has to take account of about 2^{k+1} words. Then unless p turns out to be much smaller than n, we are inevitably running here into the domain of nonpolynomial time complexity. So layers are huge while the dimensions of subspaces $\mathcal{A}_k$ are small, not exceeding n^2, and consequently most of the new words from the k-th layer added in the process of forming $\mathcal{A}_k$ will turn out linearly dependent with respect to the earlier processed ones. Yet p saturating the sequence of inclusions of $\mathcal{A}_p$ may very well be comparable with n or even worse than that. The check whether the next added word increases the dimension of $\mathcal{A}_k$ can itself be done by a Gaussian elimination algorithm at polynomial cost.

Let us analyze in turn the complexity of computing $\mathcal{N}_k$ by (11) under the assumption that a basis of $\mathcal{A}$ is given. Similarly as in the case of exterior-algebraic approach described previously, the time complexity here depends critically on k. Firstly, the number of terms in the standard polynomial Q_{2k} grows very rapidly being equal $(2k)!$. Secondly, as indicated above, the intersection in (11) has to extend over all $(2k+1)$-tuples of d basis elements of $\mathcal{A}$, where $d = \dim \mathcal{A}$. Hence the number of terms to account for is d^{2k+1}, which in the worst case of $d \sim n^2$ is of the order of $O(n^{4k+2})$. For $k = 2$ it is $O(n^{10})$. We can see again that such a direct method of construction of $\mathcal{N}_k$ can be carried out in practice only for small k. It is a separate and interesting issue to explore to what extent can prior knowledge of some properties of A and B simplify the computation of $\mathcal{N}_k$. For instance, the nondegeneracy of spectra of A or B can be expected to help.

In the discussion of consequences of Theorem 6 the following two corollaries can be immediately formulated:

1. *If W is an invariant subspace of $\mathcal{A}$ such that* $\dim W \leq k$, *then it is necessarily contained in* $\mathcal{N}_k$.
2. *$\mathcal{A}$ has a nontrivial invariant subspace with dimension not exceeding k iff* $\mathcal{N}_k \neq \{0\}$.

While this constitutes some improvement over the previous exterior-algebraic treatment of the existence of k-dimensional common invariant subspaces, the very question for a fixed value of k cannot be fully answered on the basis of Theorem 6 alone. Let us mention here only, without going into details which prove to be quite technical in this case, some more results addressing this issue. In [9] the complete solution for $k = 2$ is given and it is indicated that in the case of semisimple algebras $\mathcal{A}$ there is a complete rational solution for of the problem for any $1 < k < n$. In [3], the following theorem is proved.

THEOREM 7. *Let $\mathcal{A} = \mathcal{A}(A, B)$ be a semisimple algebra. Then $\mathcal{A}$ has an irreducible*[3] *invariant subspace of dimension k iff* $\dim \mathcal{N}_{k-1} < \dim \mathcal{N}_k$.

Moreover, this result is further extended to arbitrary algebras by means of restricting the analysis to the so-called socle of $\mathcal{A}$, which is the maximal invariant subspace Λ of $\mathcal{A}$ such that the restriction $\mathcal{A}|_\Lambda$ is a semisimple algebra. Hence one can use Theorem 7 for $\mathcal{A}|_\Lambda$. Then, since Λ can be shown to contain all irreducible invariant subspaces of $\mathcal{A}$, the solution turns out to be valid also for the original algebra $\mathcal{A}$.

[3] W is an irreducible invariant subspace of $\mathcal{A}$ if the restriction of $\mathcal{A}$ to W coincides with entire $\mathcal{M}_k$, where k is the dimension of W.

It should be noted that for a finite-dimensional algebra $\mathcal{A}$ checking it for semisimplicity as well as the construction of the socle of $\mathcal{A}$ can all be done by finite rational procedures. They can be reduced to a Gaussian elimination on a $d \times d$ matrix, where $d = \dim \mathcal{A}$. Here again we assume that some basis of $\mathcal{A}$ is given, for otherwise we run into the intractable problem of constructing it.

Finally, let us point the reader to yet another approach [4] discussing a solution of the common invariant subspace problem in the language of algebraic geometry and Gröbner bases.

5.3. The application of Burnside's theorem

Let us begin with the formulation of the theorem.

THEOREM 8 (Burnside). *Any subalgebra $\mathcal{A}$ of $\mathcal{M}_n(\mathbb{C})$ whose only invariant subspaces are $\{0\}$ and $\mathbb{C}^n$ is necessarily equal to $\mathcal{M}_n(\mathbb{C})$.*

This result can be used to verify whether a given set of operators generates the whole matrix algebra $\mathcal{M}_n$, so it has natural application in analyzing sufficiency of various sets of observables. Let us also note that the question of irreducibility of a quantum operation Φ is equivalent to saying that the collection of Kraus operators for Φ (1) generates $\mathcal{M}_n$.

When $\mathcal{A} = \mathcal{A}(A, B)$, then Shemesh criterion is the tool that can be used directly to verify the assumption in Burnside's theorem: if $\mathcal{N} = \{0\}$ then $\mathcal{A}(A, B) = \mathcal{M}_n$. Suppose in turn that the algebra $\mathcal{A}$ is generated by more than two operators, $\mathcal{A} = \mathcal{A}(A_1, \ldots, A_p)$. We can adopt the following strategy.

1. Compute Shemesh kernels $\mathcal{N}(A_i, A_j)$ for all pairs of operators.
2. Find the intersection $\Lambda_1 = \bigcap_{i,j} \mathcal{N}(A_i, A_j)$. If $\Lambda_1 = \{0\}$, then $\mathcal{A} = \mathcal{M}_n$, otherwise continue to step 3.
3. Replace the operators A_i with their restrictions to Λ_1, $A_i := A_i|_{\Lambda_1}$ and carry on steps 1 and 2 to obtain Λ_2. If $\Lambda_2 = \Lambda_1$, then Λ_2 is the nontrivial invariant subspace of $\mathcal{A}$ and consequently $\mathcal{A} \neq \mathcal{M}_n$. Otherwise iterate 3 with Λ_2 in place of Λ_1 to obtain Λ_3 and so on.

Clearly we have

$$\Lambda_1 \supset \Lambda_2 \supset \cdots \supset \{0\},$$

so either all the inclusions above are proper and after a finite number of iterations we must end up with $\Lambda_t = \{0\}$, or $\Lambda_t = \Lambda_{t+1} \neq \{0\}$ for some t. Hence this procedure terminates. Let us estimate its complexity. There are $\binom{p}{2} \sim p^2$ kernels to compute in step one, so its cost is bounded by $O(p^2 n^5)$. The construction of Λ_1 can be realized iteratively with the use of Gauss elimination at the total cost of at most $O(p^2 n^3)$. Finally, the number of iterations of step 3 is bounded by the dimension of $\mathcal{A}$, that is by n^2. Consequently the upper bound on the complexity of the entire procedure is $O(p^2 n^7)$.

6. Conclusions

We have seen that some rational computational procedures, while very useful for quantum information theoretic analyses, have nonpolynomial time complexity which in principle disqualifies them from practical applications. The polynomial complexity bounds obtained for procedures using the Shemesh criterion may also look somewhat pessimistic, yet they are certainly crude and we believe there is plenty of room for improvement if one uses some extra knowledge about the operators taking part in the computation. There is an apparent need for efficient algorithms for the construction of bases of finite-dimensional algebras — without such methods many of the procedures discussed here cannot be carried out efficiently. It is possible that some efficient Monte Carlo methods could be designed for such a class of problems. Such situation is not uncommon in computational algebra, as many of its problems belong to the BPP class. We hope to address some of these issues in future research.

Author details

Miłosz Michalski

Institute of Physics, Nicolaus Copernicus University, Toruń, Poland

References

[1] Alpin Yu. A., L. Elsner, K. D. Ikramov, Linear Algebra & Appl. **306** (2000), 165–182.

[2] Alpin Yu. A., A. George, K. D. Ikramov, Linear Algebra & Appl. **312** (2000), 115–123.

[3] Alpin Yu. A., K. D. Ikramov, J. Math. Sci. **114**:6 (2003), 1757–1764.

[4] Arapura D., C. Peterson, Linear Algebra & Appl. **384** (2004), 1–7.

[5] Chistov A., H. Fournier, L. Gurvits, P. Koiran, Found. Computational Math. **3**:4 (2003), 421–427.

[6] Choi M. D., D. W. Kribs, K. Życzkowski, Rep. Math. Phys. **58** (2006), 77–91.

[7] Cook S. A., *The complexity of theorem proving procedures*, Proc. 3rd Annual ACM Symposium on the Theory of Computing, ACM, New York, 1971, 151–158.

[8] Glynn D. G., European J. of Combinatorics **31**(2010), 1887–1891.

[9] George A., K. D. Ikramov, Linear Algebra & Appl. **287** (1999), 171–179.

[10] Gerstenhaber M., Ann. Math. **73** (1961), 324–348.

[11] Gorini V., A. Kossakowski, E. C. G. Sudarshan, J. Math. Phys. **17** (1976), 821–825.

[12] Jamiołkowski A., Rep. Math. Phys. **46**:3 (2000), 469–482.

[13] Jamiołkowski A., *On sufficient algebraic conditions for identification of quantum states*, in: "Quantum Bio-Informatics IV", L. Accardi, W. Freudenberg, M. Ohya (Eds.), World Scientific, Singapore, 2011, 185–197.

[14] Jamiołkowski A., Int. J. Geometric Methods in Modern Physics **9**:2 (2012), 1260014.

[15] Jamiołkowski A., *On applications of PI-algebras in the analysis of quantum channels*, this volume.

[16] Jamiołkowski A., private communication.

[17] Knill E., R. Laflamme, Phys. Rev. A **55** (1997), 900–911.

[18] Lidar D. A., I. L. Chuang, K. B. Whaley, Phys. Rev. Lett. **81** (1998), 2594–2597.

[19] Lidar A. A., D. Bacon, K. B. Whaley, Phys. Rev. Lett. **82** (1999), 4556–4559.

[20] Lindblad G., Comm. Math. Phys. **48** (1976), 119–130.

[21] Pappacena C. J., J Algebra **197** (1997), 535–545.

[22] Paz A., Linear & Multilinear Algebra **15** (1984), 161–170.

[23] Ryser H. J., *Combinatorial Mathematics*, The Carus Mathematical Monographs 14, Wiley, N.Y., 1965.

[24] Shemesh D., Linear Algebra Appl. **62** (1984) 11–18.

[25] Valiant L. G., Theoret. Comp. Sci. **8**:2 1979, 189–201.

[26] See e.g. `http://en.wikipedia.org/wiki/Wang_tile`

[27] See e.g. `http://en.wikipedia.org/wiki/Sylvester_matrix`

Chapter 3

Detecting Quantum Entanglement: Positive Maps and Entanglement Witnesses

Dariusz Chruściński

Additional information is available at the end of the chapter

1. Introduction

The interest on quantum entanglement has dramatically increased during the last 2 decades due to the emerging field of quantum information theory. It turns out that quantum entanglement may be used as basic resources in quantum information processing and communication. The prominent examples are quantum cryptography, quantum teleportation, quantum error correction codes, and quantum computation. Since the quantum entanglement is the basic resource for the new quantum information technologies, it is therefore clear that there is a considerable interest in efficient theoretical and experimental methods of entanglement detection.

One of the most important problems of quantum information theory [1–3] is the characterization of mixed states of composed quantum systems. In particular it is of primary importance to test whether a given quantum state exhibits quantum correlation, i.e. whether it is separable or entangled. For low-dimensional systems there exists simple necessary and sufficient condition for separability. The celebrated Peres-Horodecki criterion states that a state of a bipartite system living in $\mathbb{C}^2 \otimes \mathbb{C}^2$ or $\mathbb{C}^2 \otimes \mathbb{C}^3$ is separable iff its partial transpose is positive. Unfortunately, for higher-dimensional systems there is no single *universal* separability condition.

It turns out that the above problem may be reformulated in terms of positive linear maps in operator algebras [4]: a state ρ in $\mathcal{H}_1 \otimes \mathcal{H}_2$ is separable iff $(\mathrm{id} \otimes \varphi)\rho$ is positive for any positive map φ which sends positive operators on $\mathcal{H}_2$ into positive operators on $\mathcal{H}_1$. Therefore, a classification of positive linear maps between operator algebras $\mathcal{B}(\mathcal{H}_1)$ and $\mathcal{B}(\mathcal{H}_2)$ is of primary importance. Unfortunately, in spite of the considerable effort, the structure of positive maps is still poorly understood (see "classical" papers on positive maps [5]–[17] and some recent papers [18]–[63]).

In this paper we provide characterization of important classes of positive maps in finite dimensional matrix algebras. Equivalently, due to the Choi-Jamiołkowski isomorphism, we characterized the corresponding classes of entanglement witnesses. Concerning the application in quantum entanglement theory the key role is played by indecomposable witnesses which can detect PPT entangled states, that is, a PPT state ρ_{AB} is entangled iff there exists an indecomposable entanglement witness W such that $\mathrm{Tr}(W\rho) < 0$. We illustrate the general presentation with several examples of indecomposable positive maps/entanglement witnesses: the Choi-like maps in $M_3(\mathbb{C})$, its generalizations in $M_d(\mathbb{C})$, and the Robertson map in $M_4(\mathbb{C})$ together with its generalizations in $M_{2k}(\mathbb{C})$. These examples enables one to discuss several properties like optimality and/or exposedness which are crucial in entanglement theory.

2. Positive maps and entanglement witnesses

In this paper we restrict our analysis to linear maps

$$\Lambda : M_d(\mathbb{C}) \rightarrow M_d(\mathbb{C}) , \tag{1}$$

where $M_d(\mathbb{C})$ denotes a set of $d \times d$ complex matrices. Let $M_d(\mathbb{C})^+$ be a convex set of semi-positive elements in $M_d(\mathbb{C})$.

Definition 1. *One calls Λ a positive map if $\Lambda a \in M_d(\mathbb{C})^+$ for any $a \in M_d(\mathbb{C})^+$. Similarly, Λ is k-positive if*

$$\Lambda_{(k)} := \mathrm{id}_k \otimes \Lambda \ : M_k(\mathbb{C}) \otimes M_d(\mathbb{C}) \longrightarrow M_k(\mathbb{C}) \otimes M_d(\mathbb{C}) , \tag{2}$$

is positive. Finally, Λ is completely positive (CP) if it is k-positive for all k.

Let $\mathcal{P}_k$ denotes a convex set of k-positive maps in $M_d(\mathbb{C})$. One has $\mathcal{P}_k \subset \mathcal{P}_l$ for $k > l$. Actually, due to the Choi theorem any d-positive map in $M_d(\mathbb{C})$ is CP, and hence $\mathcal{P}_{\mathrm{CP}} = \mathcal{P}_d$. Therefore, one has the following chain of proper inclusions

$$\mathcal{P}_{\mathrm{CP}} \subset \mathcal{P}_{d-1} \subset \ldots \subset \mathcal{P}_1 , \tag{3}$$

where $\mathcal{P}_1$ denotes a set of all positive maps in $M_d(\mathbb{C})$. Let $\{e_1, \ldots, e_d\}$ denotes an orthonormal basis in $\mathbb{C}^d$, and let "T" denotes a transposition map with respect to this basis, i.e. for any $a = \sum_{ij} a_{ij} e_{ij}$ one has $T(a) = \sum_{ij} a_{ij} e_{ji}$, where $e_{ij} := |e_i\rangle\langle e_j|$.

Definition 2. *One calls a linear map Λ k-copositive if the map $\Lambda \circ \mathrm{T}$ is k-positive.*

Let $\mathcal{P}^k$ denotes a convex set of k-copositive maps. One has

$$\mathcal{P}^{\mathrm{CP}} \subset \mathcal{P}^{d-1} \subset \ldots \subset \mathcal{P}^1 , \tag{4}$$

where $\mathcal{P}^1$ denotes a set of all copositive maps in $M_d(\mathbb{C})$, and $\mathcal{P}^{\mathrm{CP}}$ stands for a set of completely copositive maps (CcP). Let $\mathcal{P}_l^k$ denotes a set of maps which are l-positive and k-copositive. One has the following relations

$$\mathcal{P}_{\mathrm{CP}}^{\mathrm{CP}} \subset \mathcal{P}_{d-1}^{d-1} \subset \ldots \subset \mathcal{P}_1^1 . \tag{5}$$

Definition 3. *A positive map* $\Lambda \in \mathcal{P}_1$ *is called* decomposable *if*

$$\Lambda = \Lambda_1 + \Lambda_2 \,, \tag{6}$$

where $\Lambda_1 \in \mathcal{P}_{\rm CP}$ *and* $\Lambda_2 \in \mathcal{P}^{\rm CP}$*. A map which is not decomposable is called* indecomposable. *A positive map* $\Lambda \in \mathcal{P}_1$ *is called* atomic *if it cannot be written as in (6), where* $\Lambda_1 \in \mathcal{P}_2$ *and* $\Lambda_2 \in \mathcal{P}^2$.

It is clear that each atomic map is indecomposable but the converse is not true. Since $\mathcal{P}_1$ is a convex set it is fully characterized by its extreme elements. Clearly a positive map Λ is *extremal* if for any $\Psi \in \mathcal{P}_1$, a map $\Lambda - \Psi$ is not positive. Finally, a positive map Λ is *optimal* if for any $\Psi \in \mathcal{P}_{\rm CP}$, a map $\Lambda - \Psi$ is not positive. It is evident that each extremal map is optimal but the converse is not true.

The vectors spaces of linear maps in $M_d(\mathbb{C})$ and linear operators in $M_d(\mathbb{C}) \otimes M_d(\mathbb{C})$ have the same dimensions d^2, and hence they are isomorphic. Fixing an orthonormal basis $\{e_1, \ldots, e_d\}$ in $\mathbb{C}^d$ one may establish the following isomorphism known in the quantum information community as the *Choi-Jamiołkowski isomorphism.*

Theorem 1. *A space of linear maps in* $M_d(\mathbb{C})$ *is isomorphic to the space of linear operators in* $M_d(\mathbb{C}) \otimes M_d(\mathbb{C})$*. The corresponding isomorphism is provided by the following formula: for a linear map* Λ *one defines a linear operator* $W_\Lambda \in M_d(\mathbb{C}) \otimes M_d(\mathbb{C})$*:*

$$W_\Lambda = ({\rm id} \otimes \Lambda) P_d^+ \,, \tag{7}$$

where P_d^+ *denotes a canonical maximally entangled state in* $\mathbb{C}^d \otimes \mathbb{C}^d$

$$P_d^+ = \frac{1}{d} \sum_{i,j=1}^{d} e_{ij} \otimes e_{ij} \,. \tag{8}$$

The inverse formula reads

$$\Lambda_W(a) = d\, {\rm Tr}_2(W \cdot [\mathbb{I}_d \otimes T(a)]) \,. \tag{9}$$

Actually, it is inner product isomorphism, that is,

$$\langle\langle \Lambda_A | \Lambda_B \rangle\rangle = \langle A | B \rangle \,, \tag{10}$$

where $\langle A|B\rangle = {\rm Tr}(A^\dagger B)$ *, and*

$$\langle\langle \Lambda_A | \Lambda_B \rangle\rangle = \sum_{\alpha=1}^{d^2} \langle \Lambda_A(f_\alpha) | \Lambda_B(f_\alpha) \rangle \,, \tag{11}$$

with f_α *being an orthonormal basis in* $M_d(\mathbb{C})$*, i.e.* $\langle f_\alpha | f_\beta \rangle = \delta_{\alpha\beta}$.

Now, if Λ is a linear map preserving hermicity, that is, $\Lambda(X^\dagger) = [\Lambda(X)]^\dagger$, then W_Λ is hermitian [8]. If Λ is a positive map, then W_Λ satisfies [9]

$$\langle \psi \otimes \phi | W | \psi \otimes \phi \rangle \geq 0 \,, \tag{12}$$

for all $\psi, \phi \in \mathbb{C}^d$. Moreover, if Λ is CP, then $W_\Lambda \geq 0$ [10], that is,

$$\langle \Psi | W | \Psi \rangle \geq 0 \,, \tag{13}$$

for all $\Psi \in \mathbb{C}^d \otimes \mathbb{C}^d$. It is clear that (13) implies (12) but not vice versa. An operator satisfying (12) is called block-positive.

Definition 4. *A block-positive but not positive operator W is called an entanglement witness.*

It is clear that due to the Choi-Jamiołkowski isomorphism one can translate all properties of linear maps into the corresponding properties of linear operators $W \in M_d(\mathbb{C}) \otimes M_d(\mathbb{C})$. In particular one has

Definition 5. *An entanglement witness W is decomposable iff*

$$W = W_1 + W_2^\Gamma \,, \tag{14}$$

where $W_1, W_2 \geq 0$ and $A^\Gamma = (\mathrm{id} \otimes T)A$ denotes partial transposition.

Let $\mathcal{D}$ be a subset of density operators of a composite quantum system living in $\mathbb{C}^n \otimes \mathbb{C}^n$ detected by a given EW W, that is, $\mathcal{D} = \{\rho \mid \mathrm{Tr}(W\rho) < 0\}$. Given two EWs W_1 and W_2 one says that W_2 is finer than W_1 if $\mathcal{D}_1 \subset \mathcal{D}_2$, that is, all states detected by W_1 are also detected by W_2. A witness W is optimal if there is no other EW which is finer than W. It means that W detects quantum entanglement in the 'optimal way'. It is clear that the knowledge of optimal EWS is crucial to classify quantum states of composite systems. One proves the following

Proposition 1. *W is an optimal EW if and only if $W - Q$ is no longer EW for arbitrary positive operator Q.*

Authors of Ref. [32] formulated the following criterion for the optimality of W.

Proposition 2. *If the set of product vectors $x \otimes y \in \mathbb{C}^n \otimes \mathbb{C}^n$ satisfying*

$$\langle x \otimes y | W | x \otimes y \rangle = 0 \,, \tag{15}$$

span the total Hilbert space $\mathbb{C}^n \otimes \mathbb{C}^n$, then W is optimal.

The further classification of entanglement witnesses would be provided in the next section.

3. States of composite quantum systems

Let $\Psi \in \mathbb{C}^d \otimes \mathbb{C}^d$ such that $\langle\Psi|\Psi\rangle = 1$, and consider the corresponding Schmidt decomposition

$$\Psi = \sum_{k=1}^{r} \mu_k \, e_k \otimes f_k \,, \tag{16}$$

where $\mu_k > 0$ and $\sum_{k=1}^{r} \mu_k^2 = 1$. In the above formula $\{e_i\}$ and $\{f_j\}$ are two mutually orthogonal normalized vectors in $\mathbb{C}^d$. One calls the number r the Schmidt rank of Ψ – $SR(\Psi)$. It is clear that $1 \leq r \leq d$. Consider now a density operator $\rho \in M_d(\mathbb{C}) \otimes M_d(\mathbb{C})$.

Definition 6. *A Schmidt number [31] of ρ – $SN(\rho)$ – is defined by*

$$SN(\rho) = \min_{p_k, \psi_k} \left\{ \max_k \, SR(\Psi_k) \right\} , \tag{17}$$

where the minimum is taken over all possible pure states decompositions

$$\rho = \sum_k p_k \, |\Psi_k\rangle\langle\Psi_k| \,, \tag{18}$$

with $p_k \geq 0$, $\sum_k p_k = 1$ and Ψ_k are normalized vectors in $\mathbb{C}^d \otimes \mathbb{C}^d$.

Let us introduce the following family of positive cones:

$$\mathrm{V}_r = \{\rho \in (M_d \otimes M_d)^+ \mid \mathrm{SN}(\rho) \leq r\} \,. \tag{19}$$

One has the following chain of inclusions

$$\mathrm{V}_1 \subset \ldots \subset \mathrm{V}_d \equiv (M_d \otimes M_d)^+ \,. \tag{20}$$

Clearly, V_1 is a cone of separable (unnormalized) states and $V_d \smallsetminus V_1$ stands for a set of entangled states. Note, that a partial transposition $(\mathrm{id} \otimes \mathrm{T})$ gives rise to another family of cones:

$$\mathrm{V}^l = (\mathrm{id} \otimes \mathrm{T})\mathrm{V}_l \,, \tag{21}$$

such that $\mathrm{V}^1 \subset \ldots \subset \mathrm{V}^d$. One has $\mathrm{V}_1 = \mathrm{V}^1$, together with the following hierarchy of inclusions:

$$\mathrm{V}_1 = \mathrm{V}_1 \cap \mathrm{V}^1 \subset \mathrm{V}_2 \cap \mathrm{V}^2 \subset \ldots \subset \mathrm{V}_d \cap \mathrm{V}^d \,. \tag{22}$$

Note, that $\mathrm{V}_d \cap \mathrm{V}^d$ is a convex set of PPT (unnormalized) states. Finally, $\mathrm{V}_r \cap \mathrm{V}^s$ is a convex subset of PPT states ρ such that $\mathrm{SN}(\rho) \leq r$ and $\mathrm{SN}(\rho^\Gamma) \leq s$.

Proposition 3. *Let* $\Lambda : M_d(\mathbb{C}) \to M_d(\mathbb{C})$ *be a linear map.* $\Lambda \in \mathcal{P}_k$ *if and only if*

$$(\mathrm{id} \otimes \Lambda) V_k \subset V_d \,. \tag{23}$$

$\Lambda \in \mathcal{P}^k$ *if and only if*

$$(\mathrm{id} \otimes \Lambda) V^k \subset V_d \,. \tag{24}$$

Finally, $\Lambda \in \mathcal{P}^k_l$ *if and only if*

$$(\mathrm{id} \otimes \Lambda) V^k \cap V_l \subset V_d \,. \tag{25}$$

Let us denote by W a space of entanglement witnesses, i.e. a space of non-positive Hermitian operators $W \in M_d \otimes M_d$ such that $\mathrm{Tr}(W\rho) \geq 0$ for all $\rho \in \mathrm{V}_1$. Define a family of subsets $\mathrm{W}_r \subset M_d \otimes M_d$:

$$\mathrm{W}_r = \{\, W \in M_d \otimes M_d \mid \mathrm{Tr}(W\rho) \geq 0 \,,\ \rho \in \mathrm{V}_r \,\} \,. \tag{26}$$

One has

$$(M_d \otimes M_d)^+ \equiv \mathrm{W}_d \subset \ldots \subset \mathrm{W}_1 \,. \tag{27}$$

Clearly, $\mathrm{W} = \mathrm{W}_1 \smallsetminus \mathrm{W}_d$. Moreover, for any $k > l$, entanglement witnesses from $\mathrm{W}_l \smallsetminus \mathrm{W}_k$ can detect entangled states from $\mathrm{V}_k \smallsetminus V_l$, i.e. states ρ with Schmidt number $l < \mathrm{SN}(\rho) \leq k$. In particular $W \in \mathrm{W}_k \smallsetminus \mathrm{W}_{k+1}$ can detect state ρ with $\mathrm{SN}(\rho) = k$.

Consider now the following class of witnesses

$$\mathrm{W}^s_r := \mathrm{W}_r + (\mathrm{id} \otimes \mathrm{T})\mathrm{W}_s \,, \tag{28}$$

that is, $W \in \mathrm{W}^s_r$ iff

$$W = P + Q^\Gamma \,, \tag{29}$$

with $P \in \mathrm{W}_r$ and $Q \in \mathrm{W}_s$. Note, that $\mathrm{Tr}(W\rho) \geq 0$ for all $\rho \in \mathrm{V}_r \cap \mathrm{V}^s$. Hence such W can detect PPT states ρ such that $\mathrm{SN}(\rho) \geq r$ or $\mathrm{SN}(\rho^\Gamma) \geq s$.

Proposition 4. *Elements from* W^d_d *are decomposable entanglement witnesses.*

It is clear that decomposable entanglement witnesses cannot detect PPT states. One has the following chain of inclusions:

$$\mathrm{W}^d_d \subset \ldots \subset \mathrm{W}^2_2 \subset \mathrm{W}^1_1 \equiv \mathrm{W} \,. \tag{30}$$

The 'weakest' entanglement can be detected by elements from $\mathrm{W}^1_1 \smallsetminus \mathrm{W}^2_2$. We shall call them *atomic entanglement witnesses.*

Let $\mathcal{P}_1^\circ$ denote a dual cone [23, 64] to the convex cone $\mathcal{P}_1$ of positive maps

$$\mathcal{P}^\circ = \mathrm{conv}\{ P_x \otimes P_y \,;\, \langle y|\Phi(P_x)|y\rangle \geq 0 \,,\, \Phi \in \mathcal{P}_1 \} \,, \tag{31}$$

where $P_x = |x\rangle\langle x|$ and $P_y = |y\rangle\langle y|$. It is clear that $\mathcal{P}_1^{\circ\circ} = \mathcal{P}_1$, that is, one may consider $\mathcal{P}_1$ as a dual cone to the convex cone of separable operators in $\mathcal{H} \otimes \mathcal{H}$. Recall that a face of $\mathcal{P}_1$ is a convex subset $F \subset \mathcal{P}_1$ such that if the convex combination $\Phi = \lambda\Phi_1 + (1-\lambda)\Phi_2$ of $\Phi_1, \Phi_2 \in \mathcal{P}_1$ belongs to F, then both $\Phi_1, \Phi_2 \in F$. If a ray $\{\lambda\Phi : \lambda > 0\}$ is a face of $\mathcal{P}_1$ then it is called an extreme ray, and we say that Φ generates an extreme ray. For simplicity we call such Φ an extremal positive map. A face F is exposed if there exists a supporting hyperplane H for a convex cone $\mathcal{P}$ such that $F = H \cap \mathcal{P}_1$.

A positive map $\Phi \in \mathcal{P}_1$ is exposed if it generates 1-dimensional exposed face. Let us denote by $\mathrm{Ext}(\mathcal{P}_1)$ the set of extremal points and $\mathrm{Exp}(\mathcal{P}_1)$ the set of exposed points of $\mathcal{P}_1$. Due to Straszewicz theorem [64] $\mathrm{Exp}(\mathcal{P}_1)$ is a dense subset of $\mathrm{Ext}(\mathcal{P}_1)$. Thus every extreme map is the limit of some sequence of exposed maps meaning that each entangled state may be detected by some exposed positive map. Hence, a knowledge of exposed maps is crucial for the full characterization of separable/entangled states of bi-partite quantum systems.

4. Choi-like maps in $M_3(\mathbb{C})$

It is well known that all positive maps $\Lambda : M_2(\mathbb{C}) \to M_2(\mathbb{C})$, $\Lambda : M_2(\mathbb{C}) \to M_3(\mathbb{C})$ and $\Lambda : M_3(\mathbb{C}) \to M_2(\mathbb{C})$ are decomposable [5, 12]. The first example of an indecomposable positive linear map in $M_3(\mathbb{C})$ was found by Choi [10]. The (normalized) Choi map reads as follows

$$\begin{aligned} \Phi_{\mathrm{C}}(e_{ii}) &= \sum_{i,j=1}^{3} A_{ij}^{\mathrm{C}} e_{jj} \,, \\ \Phi_{\mathrm{C}}(e_{ij}) &= -\frac{1}{2} e_{ij} \,, \quad i \neq j \,, \end{aligned} \tag{32}$$

where $||A_{ij}^{\mathrm{C}}||$ is the following doubly stochastic matrix:

$$A_{ij}^{\mathrm{C}} = \frac{1}{2} \begin{pmatrix} 1 & 1 & 0 \\ 0 & 1 & 1 \\ 1 & 0 & 1 \end{pmatrix} . \tag{33}$$

Let us consider a class of positive maps in $M_4(\mathbb{C})$ defined as follows [20]

$$\Phi[a,b,c] = N_{abc}(D[a,b,c] - \mathrm{id}) \,, \tag{34}$$

where $D[a,b,c]$ is a completely positive linear map defined by

$$D[a,b,c](X) = \begin{pmatrix} y_1 & 0 & 0 \\ 0 & y_2 & 0 \\ 0 & 0 & y_3 \end{pmatrix} , \quad (35)$$

and

$$\begin{aligned} y_1 &= (a+1)x_{11} + bx_{22} + cx_{33} , \\ y_2 &= cx_{11} + (a+1)x_{22} + bx_{33} , \\ y_3 &= bx_{11} + cx_{22} + (a+1)x_{33} , \end{aligned}$$

with x_{ij} being the matrix elements of $X \in M_3(\mathbb{C})$. The normalization factor $N_{abc} = (a+b+c)^{-1}$ guarantees that $\Phi[a,b,c]$ is unital, i.e. $\Phi[a,b,c](\mathbb{I}_3) = \mathbb{I}_3$. Note, that $\Phi[a,b,c]$ gives rise to the following doubly stochastic circulant matrix

$$D = N_{abc} \begin{pmatrix} a & b & c \\ c & a & b \\ b & c & a \end{pmatrix} . \quad (36)$$

This family contains well known examples of positive maps: note that $\Phi[0,1,1](X) = \frac{1}{2}(\mathrm{Tr}X\,\mathbb{I}_3 - X)$ which reproduces the reduction map. Moreover, $\Phi[1,1,0]$ and $\Phi[1,0,1]$ reproduce Choi map and its dual, respectively. One proves the following result [20]

Theorem 2. *A map $\Phi[a,b,c]$ is positive but not completely positive if and only if*

1. $0 \le a < 2$,
2. $a+b+c \ge 2$,
3. *if* $a \le 1$, *then* $bc \ge (1-a)^2$.

Moreover, being positive it is indecomposable if and only if $4bc < (2-a)^2$.

Actually, $\Phi[a,b,c]$ is indecomposable if and only if it is atomic, i.e. it cannot be decomposed into the sum of 2-positive and 2-copositive maps. The corresponding entanglement witness reads as follows

$$W[a,b,c] = N_{abc} \sum_{i,j=1}^{3} e_{ij} \otimes W_{ij} , \quad (37)$$

with

$$\begin{aligned} W_{11} &= ae_{11} + be_{22} + ce_{33} , \\ W_{22} &= ce_{11} + ae_{22} + be_{33} , \\ W_{33} &= be_{11} + ce_{22} + ae_{33} , \\ W_{ij} &= -e_{ij} , \quad i \neq j . \end{aligned}$$

In a recent paper [54] we analyzed a special case corresponding to $a+b+c=2$. It turns out [54] that $\Phi[a,b,c]$ is parameterized by the ellipse on the bc-plane. Moreover, one proves the following

Theorem 3 ([56, 57]). *A map $\Phi[a,b,c]$ with $a+b+c=2$ is optimal iff $a \leq 1$. It is indecomposable iff $a > 0$.*

Interestingly

Theorem 4 ([58]). *A map $\Phi[a,b,c]$ with $a+b+c=2$ is exposed iff $0 < a < 1$.*

Interestingly, indecomposability of these maps may be proved by using the following family of PPT states in $\mathbb{C}^3 \otimes \mathbb{C}^3$:

$$\rho_\epsilon = N_\epsilon \left(\sum_{i,j=1}^{3} e_{ij} \otimes e_{ij} + \epsilon \sum_{i=1}^{3} e_{ii} \otimes e_{i+1,i+1} + \epsilon^{-1} \sum_{i=1}^{3} e_{ii} \otimes e_{i+2,i+2} \right) , \tag{38}$$

with $\epsilon > 0$ and $N_\epsilon = [3(1+\epsilon+\epsilon^{-1})]^{-1}$. It is well known that ρ_ϵ is entangled iff $\epsilon \neq 1$.

5. Indecomposable maps in $M_d(\mathbb{C})$ — generalized Choi maps

In this section we provide several examples of positive maps in $M_d(\mathbb{C})$ which generalize Choi map in $M_3(\mathbb{C})$.

Example 1. The Choi map in $M_3(\mathbb{C})$ may be generalized to a positive map in $M_d(\mathbb{C})$ as follows [24]: let S be a unitary shift defined by:

$$S\, e_i = e_{i+1} \,, \quad i = 1, \ldots, d \,,$$

where the indices are understood mod d. One defines

$$\tau_{d,k}(X) = (d-k)\,\epsilon(X) + \sum_{i=1}^{k} \epsilon(S^i\, X\, S^{*i}) - X \,, \quad k = 0,1,2,\ldots,d-1 \,, \tag{39}$$

where $\epsilon(X)$ denotes the following projector

$$\epsilon(X) = \sum_{k=1}^{d} e_{kk} X e_{kk} \,.$$

The map $\tau_{d,0}$ defined is completely positive and the map $\tau_{d,d-1}$ reproduces the reduction map in $M_d(\mathbb{C})$ (and hence it is completely copositive). Note that $\tau_{d,k}(\mathbb{I}_d) = (d-1)\mathbb{I}_d$, and $\mathrm{Tr}\, \tau_{d,k}(X) = (d-1)\mathrm{Tr}\, X$, hence the normalized maps

$$\Phi_{d,k}(X) = \frac{1}{d-1}\, \tau_{d,k}(X) \,, \tag{40}$$

are doubly stochastic. In particular $\Phi[1,0,1] = \Phi_{3,1}$.

Example 2. A class of maps $\varphi_{\mathbf{p}}$ parameterized by $d+1$ parameters $\mathbf{p} = (p_0, p_1, \ldots, p_d)$:

$$\begin{aligned} \Phi[\mathbf{p}](e_{11}) &= p_0 e_{11} + p_d e_{dd} \,, \\ \Phi[\mathbf{p}](e_{22}) &= p_0 e_{22} + p_1 e_{11} \,, \\ &\vdots \\ \Phi[\mathbf{p}](e_{dd}) &= p_0 e_{dd} + p_{d-1} e_{d-1,d-1} \,, \\ \Phi[\mathbf{p}](e_{ij}) &= -e_{ij} \,, \quad i \neq j \,. \end{aligned} \tag{41}$$

One proves

Theorem 5 ([19, 25]). *If* $\mathbf{p} = (p_0, p_1, \ldots, p_d)$ *satisfy*

$$\begin{aligned} &a) \quad p_1, \ldots, p_d > 0 \,, \\ &b) \quad d-1 > p_0 \geq d-2 \,, \\ &c) \quad p_1 \cdot \ldots \cdot p_d \geq (d-1-p_0)^d \,, \end{aligned}$$

then $\Phi[\mathbf{p}]$ *is a positive indecomposable map.*

Actually, $\Phi[\mathbf{p}]$ is atomic, i.e. it cannot be decomposed into the sum of a 2-positive and 2-copositive maps. In particular the corresponding EW for $d = 3$ reads as follows

$$W[\mathbf{p}] = N[\mathbf{p}] \sum_{i,j=1}^{3} e_{ij} \otimes W_{ij} \,, \tag{42}$$

with

$$\begin{aligned} W_{11} &= p_0 e_{11} + p_3 e_{33} \,, \\ W_{22} &= p_0 e_{22} + p_1 e_{11} \,, \\ W_{33} &= p_0 e_{33} + p_2 e_{22} \,, \\ W_{ij} &= -e_{ij} \,, \quad i \neq j \,, \end{aligned}$$

and the normalization factor reads $N[\mathbf{p}] = (3p_0 + p_1 + p_2 + p_3)^{-1}$. In particular if $p_1 = p_2 = p_3 = c$, then $W[\mathbf{p}] = W[p_0, 0, c]$.

6. Entanglement witnesses based on spectral conditions

Any entanglement witness W can be represented as a difference $W = W_+ - W_-$, where both W_+ and W_- are semi-positive operators in $\mathbb{C}^d \otimes \mathbb{C}^d$. However, there is no general method to recognize that W defined by $W_+ - W_-$ is indeed an EW. One particular method based on spectral properties of W was presented in [42]. Let ψ_α $(\alpha = 1, \ldots, D = d^2)$ be an orthonormal basis in $\mathbb{C}^d \otimes \mathbb{C}^d$ and denote by P_α the corresponding projector $P_\alpha = |\psi_\alpha\rangle\langle\psi_\alpha|$. It leads therefore to the following spectral resolution of identity

$$\mathbb{I}_d \otimes \mathbb{I}_d = \sum_{\alpha=1}^{D} P_\alpha \, . \tag{43}$$

Having defined eigenvectors of W one needs the corresponding eigenvalues: let $\lambda_\alpha^- \leq 0$, for $\alpha = 1, \ldots, L < D$, and $\lambda_\alpha^+ > 0$ for $\alpha = L+1, \ldots, D$, that is,

$$W_- = -\sum_{\alpha=1}^{L} \lambda_\alpha^- P_\alpha \, , \quad W_+ = \sum_{\alpha=L+1}^{D} \lambda_\alpha^+ P_\alpha \, .$$

Let us analyze the condition for the spectrum $\{\lambda_\alpha^-, \lambda_\alpha^+\}$ which guarantees that W is block positive. Consider a normalized vector $\psi \in \mathbb{C}^d \otimes \mathbb{C}^d$ and let $s_1(\psi) \geq \ldots \geq s_d(\psi)$ denote its Schmidt coefficients. For any $1 \leq k \leq d$ one defines k-norm of ψ by the following formula

$$||\psi||_k^2 = \sum_{j=1}^{k} s_j^2(\psi) \, . \tag{44}$$

It is clear that $||\psi||_1 \leq ||\psi||_2 \leq \ldots \leq ||\psi||_d$. Note that $||\psi||_1$ gives the maximal Schmidt coefficient of ψ, whereas due to the normalization, $||\psi||_d^2 = \langle\psi|\psi\rangle = 1$. In particular, if ψ is maximally entangled then

$$||\psi||_k^2 = \frac{k}{d} \, . \tag{45}$$

Equivalently one may define k-norm of ψ by

$$||\psi||_k^2 = \max_{\phi} |\langle\psi|\phi\rangle|^2 \, , \tag{46}$$

where the maximum runs over all normalized vectors ϕ such that $\mathrm{SR}(\psi) \leq k$ (such ϕ is usually called k-separable). Recall that a Schmidt rank of ψ – $\mathrm{SR}(\psi)$ – is the number of non-vanishing Schmidt coefficients of ψ. One calls entanglement witness W a k-EW if $\langle\psi|W|\psi\rangle \geq 0$ for all ψ such that $\mathrm{SR}(\psi) \leq k$. One has the the following

Theorem 6 ([42]). *Let $\sum_{\alpha=1}^{L} ||\psi_\alpha||_k^2 < 1$. If the following spectral conditions are satisfied*

$$\lambda_\alpha^+ \geq \mu_k \, , \quad \alpha = L+1, \ldots, D \, , \tag{47}$$

where

$$\mu_\ell := \frac{\sum_{\alpha=1}^{L} |\lambda_\alpha^-| \, ||\psi_\alpha||_\ell^2}{1 - \sum_{\alpha=1}^{L} ||\psi_\alpha||_\ell^2} \, , \tag{48}$$

then W is an k-EW. If moreover $\sum_{\alpha=1}^{L} ||\psi_\alpha||_{k+1}^2 < 1$ *and*

$$\mu_{k+1} > \lambda_\alpha^+ \,, \quad \alpha = L+1, \ldots, D \,, \tag{49}$$

then W being k-EW is not $(k+1)$*-EW.*

Interestingly, one has the following

Theorem 7. $W = W_+ - W_-$ *is a decomposable EW.*

The proof is easy [43]: note that $W = A + B$, where

$$A = \sum_{\alpha=L+1}^{D} (\lambda_\alpha^+ - \mu_1) P_\alpha \,, \tag{50}$$

and

$$B = \mu_1 \mathbb{I}_d \otimes \mathbb{I}_d - \sum_{\alpha=1}^{L} (|\lambda_\alpha^-| + \mu_1) P_\alpha \,. \tag{51}$$

Now, since $\lambda_\alpha^+ \geq \mu_1$, for $\alpha = L+1, \ldots, D$, it is clear that $A \geq 0$. The partial transposition of B reads as follows

$$B^\Gamma = \mu_1 \mathbb{I}_d \otimes \mathbb{I}_d - \sum_{\alpha=1}^{L} (|\lambda_\alpha^-| + \mu_1) P_\alpha^\Gamma \,. \tag{52}$$

Let us recall that the spectrum of the partial transposition of rank-1 projector $|\psi\rangle\langle\psi|$ is well know: the nonvanishing eigenvalues of $|\psi\rangle\langle\psi|^\Gamma$ are given by $s_\alpha^2(\psi)$ and $\pm s_\alpha(\psi) s_\beta(\psi)$, where $s_1(\psi) \geq \ldots \geq s_d(\psi)$ are Schmidt coefficients of ψ. Therefore, the smallest eigenvalue of B^Γ (call it $b_{\min}$) satisfies

$$b_{\min} \geq \mu_1 - \sum_{\alpha=1}^{L} (|\lambda_\alpha^-| + \mu_1) ||\psi_\alpha||_1^2 \,, \tag{53}$$

and using the definition of μ_1 (cf. Eq. (48)) one gets $b_{\min} \geq 0$ which implies $B^\Gamma \geq 0$. Hence, the entanglement witness W is decomposable.

Remark 1. Interestingly, saturating the bound (47), i.e. taking

$$\lambda_\alpha^+ = \mu_1 \,, \quad \alpha = L+1, \ldots, D \,, \tag{54}$$

one has $A = 0$ and hence $W = Q^\Gamma$ with $Q = B^\Gamma \geq 0$ which shows that the corresponding positive map $\Lambda : M_d(\mathbb{C}) \to M_d(\mathbb{C})$ defined by

$$\Lambda(X) = \mathrm{Tr}_1(W \cdot X^T \otimes \mathbb{I}_d) \,, \tag{55}$$

is completely co-positive. Note that

$$\Lambda(X) = \mu_1 \mathbb{I}_d \mathrm{Tr}\, X - \sum_{\alpha=1}^{L} (\mu_1 + |\lambda_\alpha|) F_\alpha X F_\alpha^\dagger \,, \tag{56}$$

where F_α is a linear operator $F_\alpha : \mathbb{C}^d \to \mathbb{C}^d$ defined by

$$\psi_\alpha = \sum_{i=1}^{d} e_i \otimes F_\alpha e_i \,, \tag{57}$$

and $\{e_1, \ldots, e_d\}$ denotes an orthonormal basis in $\mathbb{C}^d$. In particular, if $L = 1$, i.e. there is only one negative eigenvalue, then formula (56) (up to trivial rescaling) gives

$$\Lambda(X) = \kappa\, \mathbb{I}_d \mathrm{Tr}\, X - F_1 X F_1^\dagger \,, \tag{58}$$

with

$$\kappa = \frac{\mu_1}{\mu_1 + |\lambda_1|} = ||\psi_1||_1^2 \,. \tag{59}$$

It reproduces a positive map (or equivalently an EW $W = \kappa\, \mathbb{I}_d \otimes \mathbb{I}_d - P_1$) which is known to be completely co-positive [3, 39, 43]. If ψ_1 is maximally entangled, that is, $F_1 = U/\sqrt{d}$ for some unitary $U \in U(d)$, then one finds for $\kappa = 1/d$ and the map (58) is unitary equivalent to the reduction map $\Lambda(X) = UR(X)U^\dagger$, where $R_d(X) = \mathbb{I}_d \mathrm{Tr} X - X$.

Example 3. Consider an EW corresponding to the flip operator in $d = 2$:

$$W = e_{11} \otimes e_{11} + e_{22} \otimes e_{22} + e_{12} \otimes e_{21} + e_{21} \otimes e_{12} \,. \tag{60}$$

Its spectral decomposition has the following form

$$\psi_1 = \frac{1}{\sqrt{2}}(|12\rangle - |21\rangle) \,, \quad \psi_2 = \frac{1}{\sqrt{2}}(|12\rangle + |21\rangle) \,, \quad \psi_3 = |11\rangle \,, \quad \psi_4 = |22\rangle \,.$$

together with the corresponding eigenvalues

$$-\lambda_1^- = \lambda_2^+ = \lambda_3^+ = \lambda_4^+ = 1 \,,$$

One finds $\mu_1 = 1$ and hence condition (47) is trivially satisfied $\lambda_\alpha^+ \geq \mu_1$ for $\alpha = 2, 3, 4$. We stress that our construction does not recover flip operator in $d > 2$. It has $d(d-1)/2$ negative eigenvalues. Our construction leads to at most $d - 1$ negative eigenvalues.

7. Bell-diagonal entanglement witnesses

Let us define a generalized Bell states [65] in $\mathbb{C}^d \otimes \mathbb{C}^d$

$$\psi_{mn} = (\mathbb{I}_d \otimes U_{mn})\psi_d^+ , \tag{61}$$

where U_{mn} are unitary matrices defined as follows

$$U_{mn} e_k = \lambda^{mk} e_{k+n} , \tag{62}$$

with $\lambda = e^{2\pi i/d}$. The matrices U_{mn} define an orthonormal basis in the space $M_d(\mathbb{C})$ of complex $d \times d$ matrices. One easily shows

$$\mathrm{Tr}(U_{mn} U_{rs}^\dagger) = d\, \delta_{mr} \delta_{ns} . \tag{63}$$

Some authors [66] call U_{mn} generalized spin matrices since for $d = 2$ they reproduce standard Pauli matrices:

$$U_{00} = \mathbb{I}_2 ,\ U_{01} = \sigma_1 ,\ U_{10} = i\sigma_2 ,\ U_{11} = \sigma_3 . \tag{64}$$

One calls a Hermitian operator W in $M_d(\mathbb{C}) \otimes M_d(\mathbb{C})$ *Bell diagonal* if

$$W = \sum_{m,n=0}^{d-1} p_{mn} P_{mn} , \tag{65}$$

with $p_{mn} \in \mathbb{R}$, and

$$P_{mn} = |\psi_{mn}\rangle\langle\psi_{mn}| . \tag{66}$$

Example 4. Consider the flip operator in $d = 2$. One has

$$F = P_{00} + P_{10} + P_{01} - P_{11} , \tag{67}$$

which proves that F is Bell diagonal and possesses single negative eigenvalue.

Example 5. *Consider a family $W[a,b,c]$. One finds the following spectral representation*

$$W[a,b,c] = (a-2)P_{00} + (a+1)(P_{10} + P_{20}) + b\Pi_1 + c\Pi_2 , \tag{68}$$

where

$$\Pi_m = P_{0m} + P_{1m} + P_{2m} , \tag{69}$$

which shows that $W[a,b,c]$ is Bell diagonal with a single negative eigenvalue '$a-2$'.

Example 6. Entanglement witness corresponding to the reduction map $\Lambda(X) = \mathbb{I}\mathrm{Tr}X - X$ in $M_d(\mathbb{C})$. One has

$$W = \frac{1}{d}\mathbb{I}_d \otimes \mathbb{I}_d - P_d^+ = \frac{1}{d}\sum_{k,l=0}^{d-1} P_{kl} - P_{00} , \tag{70}$$

which shows that W is Bell diagonal with a single negative eigenvalue $(1-d)/d$.

Corollary 1. *If $L < d$ and*

$$\lambda_\alpha^+ \geq \mu_1 , \quad \alpha = L, \ldots, d^2 - 1 , \tag{71}$$

with $\mu_1 = \frac{1}{d-L}\sum_{\alpha=0}^{L-1} |\lambda_\alpha^-|$, then $W = W_+ - W_-$ defines Bell diagonal entanglement witness.

8. Optimal maps in $M_{2k}(\mathbb{C})$

In this section we provide several examples of optimal indecomposable maps in $M_{2k}(\mathbb{C})$. Interestingly, some of them turn out to be extremal and even exposed [60, 61]. Consider $X \in M_{2k}(\mathbb{C}) = M_2(\mathbb{C}) \otimes M_k(\mathbb{C})$ represented as follows

$$X = \sum_{k,l=1}^{2} e_{kl} \otimes X_{kl} , \tag{72}$$

where $X_{kl} \in M_k(\mathbb{C})$. In what follows we shall use the following notation

$$X = \left(\begin{array}{c|c} X_{11} & X_{12} \\ \hline X_{21} & X_{22} \end{array} \right) , \tag{73}$$

to display the block structure of X. Robertson map [13] in $M_4(\mathbb{C})$ is defined as follows:

$$\Phi_4 \left(\begin{array}{c|c} X_{11} & X_{12} \\ \hline X_{21} & X_{22} \end{array} \right) = \frac{1}{2} \left(\begin{array}{c|c} \mathbb{I}_2 \,\mathrm{Tr}X_{22} & -[X_{12} + R_2(X_{21})] \\ \hline -[X_{21} + R_2(X_{12})] & \mathbb{I}_2 \,\mathrm{Tr}X_{11} \end{array} \right) , \tag{74}$$

where R_2 is a reduction map in $M_2(\mathbb{C})$

$$R_2(X) = \mathbb{I}_2 \mathrm{Tr}X - X . \tag{75}$$

Theorem 8. *Φ_4 defines positive indecomposable map. Moreover, it is extremal in the convex set of positive maps in $M_4(\mathbb{C})$.*

Interestingly

Theorem 9 ([61]). *Φ_4 is an exposed map.*

Following [35] and [34] one defines

$$\Phi_{2k}^{U}(X) = \frac{1}{2(k-1)}\left[R_n(X) - UX^T U^\dagger\right], \tag{76}$$

where U is an antisymmetric unitary matrix in $M_{2k}(\mathbb{C})$. The above normalization guaranties that Φ_{2k}^{U} is unital. The characteristic feature of these maps is that for any rank-1 projector P its image under Φ_{2k}^{U} reads as follows:

$$\Phi_{2k}^{U}(P) = \frac{1}{2(k-1)}\left[\mathbb{I}_{2k} - P - Q\right], \tag{77}$$

where $Q = UP^{\mathrm{T}}U^\dagger$ is a rank-1 projector orthogonal to P. Hence $\Phi_{2k}^{U}(P)$ is a projector which proves positivity of Φ_{2k}^{U}. Denote by U_0 the following "canonical" antisymmetric unitary matrix in $M_{2k}(\mathbb{C})$

$$U_0 = \mathbb{I}_k \otimes J, \tag{78}$$

where J is a symplectic matrix in $M_2(\mathbb{R})$, that is,

$$J = \begin{pmatrix} 0 & 1 \\ -1 & 0 \end{pmatrix}. \tag{79}$$

Note, that if $V \in M_{2k}(\mathbb{R})$ is orthogonal then

$$U = VU_0V^{\mathrm{T}}, \tag{80}$$

is antisymmetric and unitary. Interestingly, the map Φ_{2k}^{0} corresponding to $U = U_0$ has the following block structure

$$\Phi_{2k}^{0}\left(\begin{array}{c|c|c|c} X_{11} & X_{12} & \cdots & X_{1k} \\ \hline X_{21} & X_{22} & \cdots & X_{2k} \\ \hline \vdots & \vdots & \ddots & \vdots \\ \hline X_{k1} & X_{k2} & \cdots & X_{kk} \end{array}\right) = \frac{1}{2(k-1)}\left(\begin{array}{c|c|c|c} A_1 & -B_{12} & \cdots & -B_{1k} \\ \hline -B_{21} & A_2 & \cdots & -B_{2k} \\ \hline \vdots & \vdots & \ddots & \vdots \\ \hline -B_{k1} & -B_{k2} & \cdots & A_k \end{array}\right), \tag{81}$$

where

$$A_k = \mathbb{I}_2(\mathrm{Tr}X - \mathrm{Tr}X_{kk}), \tag{82}$$

and

$$B_{kl} - X_{kl} - R_2(X_{lk}), \tag{83}$$

and hence it reduces for $k = 2$ to the Robertson map (74).

In the recent paper [44] we proposed another construction of maps in $M_{2k}(\mathbb{C})$. Now, instead of treating a $2k \times 2k$ matrix X as a $k \times k$ matrix with 2×2 blocks X_{ij} we consider alternative possibility, i.e. we consider X as a 2×2 with $k \times k$ blocks and define

$$\Psi_{2k}\left(\begin{array}{c|c} X_{11} & X_{12} \\ \hline X_{21} & X_{22} \end{array}\right) = \frac{1}{k}\left(\begin{array}{c|c} \mathbb{I}_k \,\mathrm{Tr} X_{22} & -[X_{12} + R_k(X_{21})] \\ \hline -[X_{21} + R_k(X_{12})] & \mathbb{I}_k \,\mathrm{Tr} X_{11} \end{array}\right) . \tag{84}$$

Again, normalization factor guaranties that the map is unital, i.e. $\Psi_{2k}(\mathbb{I}_2 \otimes \mathbb{I}_k) = \mathbb{I}_2 \otimes \mathbb{I}_k$. It is clear that for $k = 2$ one has $\Psi_4 = \Phi^0_4$.

Theorem 10 ([44]). *Ψ_{2k} defines a linear positive map in $M_{2k}(\mathbb{C})$. Moreover, it is an atomic and optimal map.*

In Ref. [51] we proposed the following generalization of the Robertson map Φ_{2k}: for any collection of $k(k-1)/2$ complex numbers z_{ij}, with $i < j$, satisfying $|z_{ij}| \leq 1$ we define $\Phi^{(\mathbf{z})}_{2k} : M_{2k}(\mathbb{C}) \longrightarrow M_{2k}(\mathbb{C})$ by

$$\Phi^{(\mathbf{z})}_{2k}\left(\begin{array}{c|c|c|c} X_{11} & X_{12} & \cdots & X_{1k} \\ \hline X_{21} & X_{22} & \cdots & X_{2k} \\ \hline \vdots & \vdots & \ddots & \vdots \\ \hline X_{k1} & X_{k2} & \cdots & X_{kk} \end{array}\right) = \frac{1}{2(k-1)}\left(\begin{array}{c|c|c|c} A_1 & -z_{12}B_{12} & \cdots & -z_{1k}B_{1k} \\ \hline -\overline{z}_{21}B_{21} & A_2 & \cdots & -z_{2k}B_{2k} \\ \hline \vdots & \vdots & \ddots & \vdots \\ \hline -\overline{z}_{k1}B_{k1} & -\overline{z}_{k2}B_{k2} & \cdots & A_k \end{array}\right) . \tag{85}$$

It is clear that form $z_{ij} = 1$ the map $\Phi^{(\mathbf{z})}_{2k}$ reduces to Φ_{2k}. One proves

Theorem 11 ([51]). *$\Phi^{(\mathbf{z})}_{2k}$ defines a positive map. Moreover, $\Phi^{(\mathbf{z})}_{2k}$ is optimal and indecomposable iff* $|z_{ij}| = 1$.

9. Conclusions

We provide characterization of several classes of positive maps in $M_d(\mathbb{C})$. Equivalently, due to the Choi-Jamiołkowski isomorphism, we characterized the corresponding classes of entanglement witnesses. Concerning the application in quantum entanglement theory the key role is played by indecomposable maps which can detect PPT entangled states. The presentation was illustrated with several examples of indecomposable positive maps/entanglement witnesses: the Choi-like maps in $M_3(\mathbb{C})$ and its generalizations in $M_d(\mathbb{C})$. It was shown that several maps from these families are optimal and even exposed. Similarly, the Robertson map in $M_4(\mathbb{C})$ and its generalizations in $M_{2k}(\mathbb{C})$ turn out to be optimal maps [original Robertson map is even exposed]. It should be stressed that there is no general method enabling one to construct all indecomposable positive maps and hence this subject deserves further studies.

Acknowledgments

The author thanks Andrzej Kossakowski, Lech Woronowicz, Gniewko Sarbicki, Seung-Hyeok Kye, Justyna Pytel Zwolak, Filip Wudarski and Adam Rutkowski for many discussions about the structure of positive maps and entanglement witnesses. This work was partially supported by the National Science Center project DEC-2011/03/B/ST2/00136.

Author details

Dariusz Chruściński

Institute of Physics, Nicolaus Copernicus University, Toruń, Poland

References

[1] M. A. Nielsen and I. L. Chuang, *Quantum computation and quantum information*, Cambridge University Press, Cambridge, 2000.

[2] R. Horodecki, P. Horodecki, M. Horodecki and K. Horodecki, Rev. Mod. Phys. 81, 865 (2009).

[3] O. Gühne and G. Tóth, Phys. Rep. 474, 1 (2009).

[4] V. Paulsen, *Completely Bounded Maps and Operator Algebras*, Cambridge University Press, 2003.

[5] E. Størmer, Acta Math. 110, 233 (1963); Trans. Amer. Math. Soc. 120, 438 (1965); Proc. Am. Math. Soc. 86, 402 (1982).

[6] E. Størmer, in Lecture Notes in Physics 29, Springer Verlag, Berlin, 1974, pp. 85-106.

[7] W. Arverson, Acta Math. 123, 141 (1969).

[8] J. de Pillis, Pacific J. Math. 23, 129 (1967).

[9] A. Jamiołkowski, Rep. Math. Phys. 3, 275 (1972).

[10] M.-D. Choi, Lin. Alg. Appl. 10, 285 (1975); *ibid* 12, 95 (1975).

[11] M.-D. Choi, J. Operator Theory, 4, 271 (1980).

[12] S.L. Woronowicz, Rep. Math. Phys. 10, 165 (1976); Comm. Math. Phys. 51, 243 (1976).

[13] A.G. Robertson, J. London Math. Soc. (2) 32, 133 (1985); Quart. J. Math. Oxford (2), 34, 87 (1983); Proc. Roy. Soc. Edinburh Sect. A, 94, 71 (1983); Math. Proc. Camb. Phil. Soc., 94, 71 (1983).

[14] T. Takasaki and J. Tomiyama, Math. Japonica, 1, 129 (1982).

[15] J. Tomiyama, Contemporary Math. 62, 357 (1987).

[16] K. Tanahashi and J. Tomiyama, Canad. Math. Bull. 31, 308 (1988).

[17] W.-S. Tang, Lin. Alg. Appl. 79, 33 (1986).

[18] H. Osaka, Lin. Alg. Appl. 153, 73 (1991); *ibid* 186, 45 (1993).

[19] H. Osaka, Publ. RIMS Kyoto Univ. 28, 747 (1992).

[20] S. J. Cho, S.-H. Kye, and S.G. Lee, Lin. Alg. Appl. 171, 213 (1992).

[21] H.-J. Kim and S.-H. Kye, Bull. London Math. Soc. 26, 575 (1994).

[22] S.-H. Kye, Math. Proc. Cambridge Philos. Soc. 122, 45 (1997); Linear Alg. Appl. 362, 57 (2003).

[23] M.-H. Eom and S.-H. Kye, Math. Scand. 86, 130 (2000).

[24] K.-C. Ha, Publ. RIMS, Kyoto Univ., 34, 591 (1998);

[25] K.-C. Ha, Lin. Alg. Appl. 348, 105 (2002); *ibid* 359, 277 (2003).

[26] K.-C. Ha, S.-H. Kye and Y. S. Park, Phys. Lett. A 313, 163 (2003); Phys. Lett. A 325, 315 (2004); J. Phys. A: Math. Gen. 38, 9039 (2005).

[27] B. M. Terhal, Lin. Alg. Appl. 323, 61 (2001).

[28] W.A. Majewski and M. Marcinek, J. Phys. A: Math. Gen. 34, 5836 (2001).

[29] A. Kossakowski, Open Sys. Information Dyn. 10, 1 (2003).

[30] K. Takasaki and J. Tomiyama, Mathematische Zeitschrift 184, 101 (1983).

[31] B. Terhal and P. Horodecki, Phys. Rev. A 61, 040301 (2000); A. Sanpera, D. Bruss and M. Lewenstein, Phys. Rev. A 63, 050301(R) (2001).

[32] M. Lewenstein, B. Kraus, J. I. Cirac and P. Horodecki, Phys. Rev. A 62, 052310 (2000).

[33] A. Sanpera, D. Bruss and M. Lewenstein, Phys. Rev. A 63, 050301 (2001).

[34] W. Hall, J. Phys. A: Math. Gen. 39, (2006) 14119.

[35] H.-P. Breuer, Phys. Rev. Lett. 97, 0805001 (2006).

[36] R. Augusiak and J. Stasińska, New Journal of Physics 11, 053018 (2009).

[37] D. Chruściński and A. Kossakowski, J. Phys. A: Math. Theor. 41 (2008) 215201.

[38] D. Chruściński and A. Kossakowski, Open Systems and Inf. Dynamics, 14, 275 (2007).

[39] M. Piani and C. Mora, Phys. Rev. A 75, 012305 (2007).

[40] D. Chruściński and A. Kossakowski, Phys. Lett. A 373 (2009) 2301-2305.

[41] D. Chruściński and A. Kossakowski, J. Phys. A: Math. Theor. 41 (2008) 145301.

[42] D. Chruściński and A. Kossakowski, Comm. Math. Phys. 290, 1051 (2009).

[43] D. Chruściński, A. Kossakowski and G. Sarbicki, Phys. Rev. A 80 (2009) 042314.

[44] D. Chruściński, J. Pytel and G. Sarbicki, Phys. Rev. A 80 (2009) 062314.

[45] D. Chruściński and J. Pytel, Phys. Rev. A 82 052310 (2010).

[46] E. Størmer, J. Funct. Anal. 254, 2303 (2008).

[47] G. Sarbicki, J. Phys. A: Math. Theor. 41, 375303 (2008).

[48] Ł. Skowronek and K. Życzkowski, J. Phys. A: Math. Theor. 42, (2009) 325302.

[49] Ł. Skowronek and K. Życzkowski, J. Math. Phys. 50, 062106 (2009).

[50] D. Chruściński, A. Kossakowski, T. Matsuoka, K. Młodawski, Open Systems and Inf. Dyn. 17, 235-254 (2010).

[51] D. Chruściński and J. Pytel, J. Phys. A: Math. Theor. 44, 165304 (2011).

[52] J. Pytel Zwolak and D. Chruściński, *New tools for investigating positive maps in matrix algebras*, arXiv:1204.6579 (to apper in Rep. Math. Phys.)

[53] D. Chruściński and A. Rutkowski, Eur. Phys. J. D 62, 273 (2011).

[54] D. Chruściński and F. A. Wudarski, Open Sys. Information Dyn. 18, 387 (2011).

[55] D. Chruściński and F. A. Wudarski, Open Sys. Information Dyn. 19, 1250020 (2012).

[56] D. Chruściński and G. Sarbicki, *Optimal entanglement witnesses for two qutrits*, arXiv:1105.4821.

[57] K-C. Ha and S-H. Kye, Phys. Rev. A 84, 024302 (2011).

[58] K-C. Ha and S-H. Kye, Open Sys. Information Dyn., 18, 323-337 (2011).

[59] R. Augusiak, G. Sarbicki, and M. Lewenstein, Phys. Rev. A 84, 052323 (2011) .

[60] D. Chruściński and G. Sarbicki, J. Phys. A: Math. Theor. 45, 115304 (2012).

[61] G. Sarbicki and D. Chruściński, *A class of exposed indecomposable positive maps*, arXiv:1201.5995 (to be publish in J. Phys. A).

[62] C. Spengler, M. Huber, S. Brierley, T. Adaktylos, and B. C. Hiesmayr, Phys. Rev. A 86, 022311 (2012).

[63] K-C. Ha and S-H. Kye, J. Phys. A: Math. and Teor. 45, 415305 (2012).

[64] R. T. Rockafellar, *Convex Analysis*, Princeton University Press, 1970.

[65] B. Baumgartner, B. Hiesmayr, and H. Narnhofer, Phys. Rev. A 74, 032327 (2006); J. Phys. A: Math. Theor. 40, 7919 (2007); Phys. Lett. A 372, 2190 (2008).

[66] O. Pittenger, M.H. Rubin, Lin. Alg. Appl. 390, 255 (2004); D. Chruściński and A. Pittenger, J. Phys. A: Math. Theor. 41, 385301 (2008).

Chapter 4

Quantum Communication Processes and Their Complexity

Noboru Watanabe

Additional information is available at the end of the chapter

1. Introduction

The complex systems and their dynamics are treated various way. Ohya looked for synthesizing method to treat complex systems. He established Information Dynamics [36] which is a new concept unifying the dynamics of a state and the complexity of the system itself. By applying Information Dynamics, one can discuss in a unified frame the problems such as in mathematics, physics, biology, information science. Information Dynamics is growing as one of the research fields, for instance, the international journal named "Open Systems and Information Dynamics" in 1992 has appeared. In ID, there are two types of complexity, that is, (a) a complexity of state describing system itself and (b) a transmitted complexity between two systems. Entropies of classical and quantum information theory are the example of the complexities of (a) and (b).

Shannon [52] found that the entropy, introduced in physical systems by Clausius and Boltzmann, can be used to express the amount of information by means of communication processes, and he proposed the so-called information communication theory at the middle part of the 20th century. In his information theory, the entropy and the mutual entropy (information) are most important concepts. The entropy relates to the complexity of ID measuring the amount of information of the state of system. The mutual entropy (information) corresponds to the transmitted complexity of ID representing the amount of information correctly transmitted from the initial system to the final system through a channel, and it was extended to the mutual entropy on the continuous probability space by Gelfand– Kolmogorov - Yaglom [17,23], which was defined by using the relative entropy of two states by Kullback-Leibler [26].

Laser is often used in the current communication. A formulation of information theory being able to treat quantum effects is necessary, which is the so-called quantum information theory. The quantum information theory is important in both mathematics and engineering. It has been developed with quantum entropy theory and quantum probability. In quantum information theory, one of the important problems is to investigate how much information is exactly transmitted to the output system from the input system through a quantum channel. The amount of information of the quantum input system is described by the quantum entropy defined by von Neumann [29] in 1932. The C*-entropy was introduced in [33,35] and its property is discussed in [28,21]. The quantum relative entropy was introduced by Umegaki [55] and it is extended to general quantum system by Araki [4,5], Uhlmann [54] and Donald [14]. Furthermore, it had been required to extend the Shannon's mutual entropy (information) of classical information theory to that in the quantum one. The classical mutual entropy is defined by using the joint probability expressing a correlation between the input system and the output system. However, it was shown by Urbanik [56] that in quantum system there does not generally exists a joint probability distribution. The semi-classical mutual entropy was introduced by Holevo, Livitin, Ingarden [18,20] for classical input and output passing through a possible quantum channel. By introducing a new notion, the so-called compound state, in 1983 Ohya formulated the mutual entropy [31,32] in a complete quantum mechanical system (i.e., input state, output state and channel are all quantum mechanical), which is called the Ohya mutual entropy. It was generalized to C*-algebra in [Oent84]. The quantum capacity [40] is defined by taking the supremum for the Ohya mutual entropy. By using the Ohya quantum mutual entropy, one can discuss the efficiency of the information transmission in quantum systems [28,27,44,34,35], which allows the detailed analysis of optical communication processes. Concerning quantum communication processes, several studies have been done in [31,32,35,40,41]. Recently, several mutual entropy type measures (Lindblad - Nielsen entropy [10] and Coherent entropy [6]) were defined by using the entropy exchange. One can classify these mutual entropy type measures by calculating their measures for the quantum channel. These entropy type complexities are explained in [39,43].

The entangled state is an important concept for quantum theory and it has been studied recently by several authors. One of the remarkable formulations to search the entanglement state is the Jamiolkowski's isomorphism [22]. It might be related to the construction of the compound state in quantum communication processes. One can discuss the entangled state generated by the beam splitting and the squeezed state.

The classical dynamical (or Kolmogorov-Sinai) entropy S(T) [23] for a measure preserving transformation T was defined on a message space through finite partitions of the measurable space. The classical coding theorems of Shannon are important tools to analyze communication processes which have been formulated by the mean dynamical entropy and the mean dynamical mutual entropy. The mean dynamical entropy represents the amount of information per one letter of a signal sequence sent from the input source, and the mean dynamical mutual entropy does the amount of information per one letter of the signal received in the output system. In this chapter, we will discuss the complexity of the quantum dynamical system to calculate the mean mutual entropy with respect to the modulated initial states and the attenuation channel for the quantum dynamical systems [59].

The quantum dynamical entropy (QDE) was studied by Connes-Størmer [13], Emch [15], Connes-Narnhofer-Thirring [12], Alicki-Fannes [3], and others [9,48,19,57,11]. Their dynamical entropies were defined in the observable spaces. Recently, the quantum dynamical entropy and the quantum dynamical mutual entropy were studied by the present authors [34,35]: (1) the dynamical entropy is defined in the state spaces through the complexity of Information

Dynamics [36]. (2) It is defined through the quantum Markov chain (QMC) was done in [2]. (3) The dynamical entropy for a completely positive (CP) map was defined in [25]. In this chapter, we will discuss the complexity of the quantum dynamical process to calculate the generalized AOW entropy given by KOW entropy for the noisy optical channel [58].

2. Quantum channels

The signal of the input quantum system is transmitted through a physical device, which is called a quantum channel. The concept of channel has been performed an important role in the progress of the quantum information communication theory. The mathematical representation of the quantum channel is a mapping from the input state space to the output state space. In particular, the attenuation channel [31] and the noisy optical channel [44] are remarkable examples of the quantum channels describing the quantum optical communication processes. These channels are related to the mathematical desctiption of the beam splitter.

Here we review the definition of the quantum channels.

Let $(B(\mathcal{H}_1), \mathfrak{S}(\mathcal{H}_1))$ and $(B(\mathcal{H}_2), \mathfrak{S}(\mathcal{H}_2))$ be input and output systems, respectively, where $B(\mathcal{H}_k)$ is the set of all bounded linear operators on a separable Hilbert space $\mathcal{H}_k$ and $\mathfrak{S}(\mathcal{H}_k)$ is the set of all density operators on $\mathcal{H}_k$ $(k=1, 2)$. Quantum channel Λ^* is a mapping from $\mathfrak{S}(\mathcal{H}_1)$ to $\mathfrak{S}(\mathcal{H}_2)$.

1. Λ^* is called a linear channel if Λ^* satisfies $\Lambda^*(\lambda\rho_1+(1-\lambda)\rho_2)=\lambda\Lambda^*(\rho_1)+(1-\lambda)\Lambda^*(\rho_2)$ for any $\rho_1, \rho_2 \in \mathfrak{S}(\mathcal{H}_1)$ and any $\lambda \in [0, 1]$.

2. Λ^* is called a completely positive (CP) channel if Λ^* is linear and its dual map Λ from $B(\mathcal{H}_2)$ to $B(\mathcal{H}_1)$ holds

$$\sum_{i,j=1}^{n} A_i^*\Lambda(B_i^* B_j)A_j \geq 0$$

for any $n \in N$, any $B_j \in B(\mathcal{H}_2)$ and any $A_j \in B(\mathcal{H}_1)$, where the dual map Λ of Λ^* is defined by $tr\Lambda^*(\rho)B = tr\rho\Lambda(B)$ for any $\rho \in \mathfrak{S}(\mathcal{H}_1)$ and any $B \in B(\mathcal{H}_2)$. Almost all physical transformations can be described by the CP channel [30,39,21,46, 43].

3. Quantum communication processes

Let $\mathcal{K}_1$ and $\mathcal{K}_2$ be two Hilbert spaces expressing noise and loss systems, respectively. Quantum communication process including the influence of noise and loss is denoted by the following scheme [31]: Let ρ be an input state in $\mathfrak{S}(\mathcal{H}_1)$, ζ be a noise state in $\mathfrak{S}(\mathcal{K}_1)$.

$$\begin{array}{ccc} \mathfrak{S}(\mathcal{H}_1) & \xrightarrow{\Lambda^*} & \mathfrak{S}(\mathcal{H}_2) \\ \gamma^* \downarrow & & \uparrow a^* \\ \mathfrak{S}(\mathcal{H}_1 \otimes \mathcal{K}_1) & \overleftrightarrow{\Pi^*} & \mathfrak{S}(\mathcal{H}_2 \otimes \mathcal{K}_2) \end{array}$$

The above maps γ^*, a^* are given as

$$\gamma^*(\rho) = \rho \otimes \xi, \quad \rho \in \mathfrak{S}(\mathcal{H}_1),$$
$$a^*(\sigma) = tr_{\mathcal{K}_2}\sigma, \quad \sigma \in \mathfrak{S}(\mathcal{H}_2 \otimes \mathcal{K}_2).$$

The map Π^* is a CP channel from $\mathfrak{S}(\mathcal{H}_1 \otimes \mathcal{K}_1)$ to $\mathfrak{S}(\mathcal{H}_2 \otimes \mathcal{K}_2)$ given by physical properties of the device transmitting signals. Hence the channel for the above process is given as

$$\Lambda^*(\rho) \equiv tr_{\mathcal{K}_2}\Pi^*(\rho \otimes \zeta) = (a^* \circ \Pi^* \circ \gamma^*)(\rho)$$

for any $\rho \in \mathfrak{S}(\mathcal{H}_1)$. Based on this scheme, the noisy optical channel is constructed as follows:

4. Noisy optical channel

Noisy optical channel Λ^* with a noise state ζ was defined by Ohya and NW [44] such as

$$\Lambda^*(\rho) \equiv tr_{\mathcal{K}_2}\Pi^*(\rho \otimes \zeta) = tr_{\mathcal{K}_2}V(\rho \otimes \zeta)V^*,$$

where $\zeta = |m_1\rangle\langle m_1|$ is the m_1 photon number state in $\mathfrak{S}(\mathcal{K}_1)$ and V is a mapping from $\mathcal{H}_1 \otimes \mathcal{K}_1$ to $\mathcal{H}_2 \otimes \mathcal{K}_2$ denoted by

$$V(|n_1\rangle \otimes |m_1\rangle) = \sum_{j=0}^{n_1+m_1} C_j^{n_1,m_1} |j\rangle \otimes |n_1+m_1-j\rangle,$$

where

$$C_j^{n_1,m_1} = \sum_{r=L}^{K} (-1)^{n_1+j-r} \frac{\sqrt{n_1!m_1!j!(n_1+m_1-j)!}}{r!(n_1-j)!(j-r)!(m_1-j+r)!} \alpha^{m_1-j+2r}(-\beta)^{n_1+j-2r},$$

and $|n_1\rangle$ is the n_1 photon number state vector in $\mathcal{H}_1$, and α, β are complex numbers satisfying $|\alpha|^2 + |\beta|^2 = 1$. K and L are constants given by $K = \min\{n_1, j\}$, $L = \max\{m_1 - j, 0\}$. We have the following theorem.

Theorem The noisy optical channel Λ^* with noise state $|m\rangle\langle m|$ is described by

$$\Lambda^*(\rho) = \sum_{i=0}^{\infty} O_i V Q^{(m)} \rho Q^{(m)*} V^* O_i^*,$$

where $Q^{(m)} \equiv \sum_{l=0}^{\infty} (|y_l\rangle \otimes |m\rangle)\langle y_l|$, $O_i \equiv \sum_{k=0}^{\infty} |z_k\rangle(\langle z_k| \otimes \langle i|)$, $\{|y_l\rangle\}$ is a CONS in $\mathcal{H}_1$, $\{|z_k\rangle\}$ is a CONS in $\mathcal{H}_2$ and $\{|i\rangle\}$ is the set of number states in $\mathcal{K}_2$.

In particular for the coherent input states

$$\rho = |\xi\rangle\langle\xi| \otimes |\kappa\rangle\langle\kappa| \in \mathfrak{S}(\mathcal{H}_1 \otimes \mathcal{K}_1),$$

the output state of Π^* is obtained by

$$\Pi^*(|\xi\rangle\langle\xi| \otimes |\kappa\rangle\langle\kappa|) = |\alpha\xi + \beta\kappa\rangle\langle\alpha\xi + \beta\kappa| \otimes |-\bar{\beta}\xi + \bar{\alpha}\kappa\rangle\langle-\bar{\beta}\xi + \bar{\alpha}\kappa|.$$

$$\begin{array}{ccccc} & & |\kappa\rangle\langle\kappa| & & \\ & & \downarrow & & \\ |\xi\rangle\langle\xi| & \to & \searrow & \to & |\alpha\xi + \beta\kappa\rangle\langle\alpha\xi + \beta\kappa| \\ & & \downarrow & & \\ & & |-\bar{\beta}\xi + \bar{\alpha}\kappa\rangle\langle-\bar{\beta}\xi + \bar{\alpha}\kappa| & & \end{array}$$

5. Attenuation channel

The noisy optical channel with a vacuum noise is called the attenuation channel introduced in [31] by

$$\Lambda_0^*(\rho) \equiv tr_{\mathcal{K}_2}\Pi_0^*(\rho \otimes \zeta_0) = tr_{\mathcal{K}_2} V_0(\rho \otimes |0\rangle\langle 0|)V_0^*,$$

where $|0\rangle\langle 0|$ is the vacuum state in $\mathfrak{S}(\mathcal{K}_1)$ and V_0 is a mapping from $\mathcal{H}_1 \otimes \mathcal{K}_1$ to $\mathcal{H}_2 \otimes \mathcal{K}_2$ given by

$$V_0(|n_1\rangle \otimes |0\rangle) = \sum_{j}^{n_1} C_j^{n_1} |j\rangle \otimes |n_1 - j\rangle,$$

$$C_j^{n_1} = \sqrt{\frac{n_1!}{j!(n_1-j)!}}\alpha^{j}(-\bar{\beta})^{n_1-j}$$

In particular, for the coherent input state

$$\rho = |\xi\rangle\langle\xi| \otimes |0\rangle\langle 0| \in \mathfrak{S}(\mathcal{H}_1 \otimes \mathcal{K}_1),$$

one can obtain the output state

$$\Pi_0^*(|\xi\rangle\langle\xi| \otimes |0\rangle\langle 0|) = |\alpha\xi\rangle\langle\alpha\xi| \otimes |-\bar{\beta}\xi\rangle\langle-\bar{\beta}\xi|.$$

Lifting $\mathcal{E}_0^*$ from $\mathfrak{S}(\mathcal{H})$ to $\mathfrak{S}(\mathcal{H} \otimes \mathcal{K})$ in the sense of Accardi and Ohya [1] is denoted by

$$\mathcal{E}_0^*(|\xi\rangle\langle\xi|) = |\alpha\xi\rangle\langle\alpha\xi| \otimes |\beta\xi\rangle\langle\beta\xi|.$$

$\mathcal{E}_0^*$ (or Π_0^*) is called a beam splitting. Based on liftings, the beam splitting was studied by Accardi - Ohya [1] and Fichtner - Freudenberg - Libsher [16].

6. Information dynamics

We are interested to study the dynamics of state change or the complexity of state for several systems. Information dynamics (ID) is a new concept introduced by Ohya [36] to construct a theory under the ID's framework by synthesizing these investigating schemes. In ID, a complexity of state describing system itself and a transmitted complexity between two systems are used. The examples of these complexities are the Shannon's entropy and the mutual entropy (information) in classical entropy theory. In quantum entropy theory, it was known that the von Neumann entropy and the Ohya mutual entropy relate to these complexities. Recently, several mutual entropy type measures (the Lindblad - Nielsen entropy [10] and the Coherent entropy [6]) were proposed by means of the entropy exchange for an input state and a channel.

7. Concept of information dynamics

Ohya introduced Information Dynamics (ID) synthesizing dynamics of state change and complexity of state. Based on ID, one can study various problems of physics and other fields. Channel and two complexities are key concepts of ID. Two kinds of complexities $C^{\mathcal{S}}(\rho)$, $T^{\mathcal{S}}(\rho;\Lambda^*)$ are used in ID. $C^{\mathcal{S}}(\rho)$is a complexity of a state ρ measured from a subset $\mathcal{S}$ and $T^{\mathcal{S}}(\rho;\Lambda^*)$ is a transmitted complexity according to the state change from ρ to $\Lambda^*\rho$. Let $\mathcal{S}$, $\bar{\mathcal{S}}$, $\mathcal{S}_t$ be subsets of $\mathfrak{S}(\mathcal{H}_1)$, $\mathfrak{S}(\mathcal{H}_2)$, $\mathfrak{S}(\mathcal{H}_1 \otimes \mathcal{H}_2)$, respectively. These complexities should fulfill the following conditions as follows:

8. Complexity of system

1. For any $\rho \in \mathcal{S}$, $C^{\mathcal{S}}(\rho)$ is nonnegative (i.e., $C^{\mathcal{S}}(\rho) \geq 0$)
2. For a bijection j from $ex\mathfrak{S}(\mathcal{H}_1)$ to $ex\mathfrak{S}(\mathcal{H}_1)$,

$$C^{\mathcal{S}}(\rho)=C^{\mathcal{S}}(j(\rho))$$

is hold, where $ex\mathfrak{S}(\mathcal{H}_1)$ is the set of all extremal points of $\mathfrak{S}(\mathcal{H}_1)$.

3. For $\rho \otimes \sigma \in \mathfrak{S}(\mathcal{H}_1 \otimes \mathcal{H}_2)$, $\rho \in \mathfrak{S}(\mathcal{H}_1)$, $\sigma \in \mathfrak{S}(\mathcal{H}_2)$,

$$C^{\mathcal{S}_t}(\rho \otimes \sigma)=C^{\mathcal{S}}(\rho)+C^{\bar{\mathcal{S}}}(\sigma)$$

It means that the complexity of the state $\rho \otimes \sigma$ of totally independent systems are given by the sum of the complexities of the states ρ and σ.

9. Transmitted complexity

(1′) For any $\rho \in S$ and a channel Λ^*, $T^S(\rho;\Lambda^*)$ is nonnegative (i.e., $T^S(\rho;\Lambda^*) \geq 0$)

(4) $C^S(\rho)$ and $T^S(\rho;\Lambda^*)$ satisfy the following inequality $0 \leq T^S(\rho;\Lambda^*) \leq C^S(\rho)$.

(5) If the channel Λ^* is given by the identity map id, then $T^S(\rho;id) = C^S(\rho)$ is hold.

The examples of the above complexities are the Shannon entropy $S(p)$ for $C^S(p)$ and the classical mutual entropy $I(p;\Lambda^*)$ for $T^S(p;\Lambda^*)$, respectively Let us consider these complexities for quantum communication processes.

10. Quantum entropy

Since the present optical communication is using the optical signal including quantum effect, it is necessary to construct new information theory dealing with those quantum phenomena in order to discuss the efficiency of information transmission of optical communication processes rigorously. The quantum information theory is important in both mathematics and engineering, and it contains several topics, for instance, quantum entropy theory, quantum communication theory, quantum teleportation, quantum entanglement, quantum algorithm, quantum coding theory and so on. It has been developed with quantum entropy theory and quantum probability. In quantum information theory, one of the important problems is to investigate how much information is exactly transmitted to the output system from the input system through a quantum channel. The amount of information of the quantum communication system is described by the quantum mutual entropy defined by Ohya [31], based on the quantum entropy by von Neumann [29], and the quantum relative entropy by Umegaki [55], Araki [4] and Uhlmann [54]. The quantum information theory directly relates to quantum communication theory, for instance, [40,41,45]. One of the most important communication processes is quantum teleportation, whose new treatment was studied in [24]. It is important to classify quantum states. One of such classifications is to study entanglement and separability of states (see [7,8]). There have been lots of trials in finite dimensional Hilbert spaces. Quantum mechanics should be basically discussed in infinite dimensional Hilbert spaces. We have to study such a classification in infinite dimensional Hilbert spaces.

10.1. Von Neumann entropy

The study of the entropy in quantum system was begun by von Neumann [29] in 1932. For any state given by the density operator ρ, the von Neumann entropy is defined by

$$S(\rho) = -tr\rho\log\rho, \quad \forall \rho \in \mathfrak{S}(\mathcal{H}).$$

Since the von Neumann entropy satisfies the conditions (1),(2),(3) of the complexity of state of ID, it seems to be considered as an example of the complexity of state $C(\rho)=S(\rho)$.

10.2. Entropy for general systems

Here we briefly explain Let us comment general entropies of states in C*-dynamical systems. The C*-entropy ($\mathcal{S}$-mixing entropy) was introduced by Ohya in [33,35] and its property is discussed in [28,21].

Let $(\mathcal{A}, \mathfrak{S}(\mathcal{A}), \alpha(G))$ be a C*-dynamical system and $\mathcal{S}$ be a weak* compact and convex subset of $\mathfrak{S}(\mathcal{A})$. For example, $\mathcal{S}$ is given by $\mathfrak{S}(\mathcal{A})$ (the set of all states on $\mathcal{A}$), $I(\alpha)$ (the set of all invariant states for α), $K(\alpha)$ (the set of all KMS states),and so on. Every state $\varphi \in \mathcal{S}$ has a maximal measure μ pseudosupported on $ex\mathcal{S}$ such that

$$\varphi=\int_{\mathcal{S}} \omega d\mu,$$

where $ex\mathcal{S}$ is the set of all extreme points of $\mathcal{S}$. The measure μ giving the above decomposition is not unique unless $\mathcal{S}$ is a Choquet simplex. The set of all such measures is denoted by $M_{\varphi}(\mathcal{S})$ and $D_{\varphi}(\mathcal{S})$is the subset of $M_{\varphi}(\mathcal{S})$ constituted by

$$D(\mathcal{S})=\Big\{M_{\varphi}(\mathcal{S});\quad \exists \mu_k \subset \mathbb{R}^+ \text{ and } \{\varphi_k\} \subset ex\mathcal{S} \quad s.t. \ \sum_k \mu_k=1, \quad \mu=\sum_k \mu_k \delta(\varphi_k)\Big\}$$

where $\delta(\varphi)$ is the Dirac measure concentrated on an initial state φ. For a measure $\mu \in D_{\varphi}(\mathcal{S})$, the entropy type functional $H(\mu)$ is given by

$$H(\mu)=-\sum_k \mu_k \log \mu_k.$$

For a state $\varphi \in \mathcal{S}$ with respect to $\mathcal{S}$, Ohya introduced the C*-entropy ($\mathcal{S}$-mixing entropy) [33,35] defined by

$$S^{\mathcal{S}}(\varphi)=\begin{cases} \inf\{H(\mu); \quad \mu \in D_{\varphi}(\mathcal{S})\} \\ +\infty \qquad\qquad \text{if } D_{\varphi}(\mathcal{S})=\varnothing. \end{cases}$$

It describes the amount of information of the state φ measured from the subsystem $\mathcal{S}$. If $\mathcal{S}=\mathfrak{S}(\mathcal{A})$, then $S^{\mathfrak{S}(\mathcal{A})}(\varphi)$ is denoted by $S(\varphi)$. This entropy is an extension of the von Neumann entropy mentioned above.

10.3. Quantum relative entropy

The classical relative entropy in continuous probability space was defined by Kullback-Leibler [26]. It was developed in noncommutative probability space. The quantum relative entropy was first defined by Umegaki [55] for σ-finite von Neumann algebras, which denotes a certain difference between two states. It was extended by Araki [4] and Uhlmann [54] for general von Neumann algebras and *-algebras, respectively.

10.4. Umegakirelative entropy

The relative entropy of two states was introduced by Umegaki in [55] for σ-finite and semi-finite von Neumann algebras. Corresponding to the classical relative entropy, for two density operators ρ and σ, it is defined as

$$S(\rho,\sigma)=\begin{cases} tr\rho(\log\rho-\log\sigma) & (s(\rho)\ll s(\sigma)),\\ \infty & (\text{else}),\end{cases}$$

where $s(\rho)\ll s(\sigma)$ means the support projection $s(\sigma)$ of σ is greater than the support projection $s(\rho)$ of ρ. It means a certain difference between two quantum states ρ and σ. The Umegaki's relative entropy satisfies (1) positivity, (2) joint convexity, (3) symmetry, (4) additivity, (5) lower semicontinuity, (6) monotonicity. Araki [4] and Uhlmann [54] extended this relative entropy for more general quantum systems.

10.5. Relative entropy for general systems

The relative entropy for two general states was introduced by Araki [4,5] in von Neumann algebra and Uhlmann [54] in *-algebra. The above properties are held for these relative entropies.

10.5.1. Araki's relative entropy[4,5]

Let $\mathcal{N}$ be a σ-finite von Neumann algebra acting on a Hilbert space $\mathcal{H}$ and φ, ψ be normal states on $\mathcal{N}$ given by $\varphi(\cdot)=\langle x,\ \cdot x\rangle$ and $\psi(\cdot)=\langle y,\ \cdot y\rangle$ with $x,\ y\in\mathcal{K}$ (i.e., $\mathcal{K}$ is a positive natural cone) $\subset\mathcal{H}$. On the domain $\mathcal{N}y+(I-s^{\mathcal{N}'}(y))\mathcal{H}$, the operator $S_{x,y}$ is defined by

$$S_{x,y}(Ay+z)=s^{\mathcal{N}}(y)A^*x,\ \ A\in\mathcal{N}\ \left(z\in\mathcal{H}\text{ is satisfying } s^{\mathcal{N}'}(y)z=0\right),$$

where $s^{\mathcal{N}}(y)$ (the $\mathcal{N}$-support of y) is the projection from $\mathcal{H}$ to $\{\mathcal{N}'y\}^-$. Using this $S_{x,y}$, the relative modular operator $\Delta_{x,y}$ is defined as $\Delta_{x,y}=(S_{x,y})^*\overline{S_{x,y}}$, whose spectral decomposition is denoted by $\int_0^\infty \lambda de_{x,y}(\lambda)$ ($\overline{S_{x,y}}$ is the closure of $S_{x,y}$). Then the Araki's relative entropy is given by

Definition The Araki's relative entropy of φ and ψ is defined by

$$S(\psi,\varphi)=\begin{cases}\int_0^\infty \log\lambda d\langle y,\ e_{x,y}(\lambda)y\rangle & (\psi\ll\varphi),\\ \infty & (\text{otherwise}),\end{cases}$$

where $\psi\ll\varphi$ means that $\varphi(A^*A)=0$ implies $\psi(A^*A)=0$ for $A\in\mathcal{N}$.

10.5.2. Uhlmann's relative entropy[54]

Let $\mathcal{L}$ be a complex linear space and $p,\ q$ be two semi-norms on $\mathcal{L}$. $H(\mathcal{L}(p,\ q))$ is the set of all positive Hermitian forms α on $\mathcal{L}$ satisfying $|\alpha(x,\ y)|\le p(x)q(y)$ for all $x,\ y\in\mathcal{L}$. For $x\in\mathcal{L}$, the quadratical mean $QM(p,\ q)$ of p and q is defined by

$QM(p, q)(x)=\sup\{\alpha(x, x)^{1/2}; \alpha \in H(\mathcal{L}(p, q))\}$.

For each $x \in \mathcal{L}$, there exists a family of semi-norms $p_t(x)$ of $t \in [0, 1]$, which is called the quadratical interpolation from p to q, satisfying the following conditions:

1. For any $x \in \mathcal{L}$, $p_t(x)$ is continuous in t,
2. $p_{1/2}=QM(p, q)$
3. $p_{t/2}=QM(p, p_t) \quad (\forall t \in [0, 1])$
4. $p_{(t+1)/2}=QM(p_t, q) \quad (\forall t \in [0, 1])$

This semi-norm p_t is denoted by $QI_t(p, q)$. It is shown that for any positive Hermitianforms α, β, there exists a unique function $QF_t(\alpha, \beta)$ of $t \in [0, 1]$ with values in the set $H(\mathcal{L}(p, q))$ such that $QF_t(\alpha, \beta)(x, x)^{1/2}$ is the quadratical interpolation from $\alpha(x, x)^{1/2}$ to $\beta(x, x)^{1/2}$. For $x \in \mathcal{L}$, the relative entropy functional $S(\alpha, \beta)(x)$ of α and β is defined as

$$S(\alpha, \beta)(x)=-\liminf_{t\to+0}\frac{1}{t}\{QF_t(\alpha, \beta)(x, x)-\alpha(x, x)\}.$$

Let $\mathcal{L}$ be a *-algebra $\mathcal{A}$. For positive linear functional φ, ψ on $\mathcal{A}$, two Hermitian forms φ^L, ψ^R are given by $\varphi^L(A, B)=\varphi(A^*B)$ and $\psi^R(A, B)=\psi(BA^*)$.

Definition The Uhlmann's relative entropy of φ and ψ is defined by

$$S(\psi, \varphi)=S(\psi^R, \varphi^L)(I).$$

10.5.3. Ohya mutual entropy [31]

The Ohya mutual entropy [31] with respect to the initial state ρ and a quantum channel Λ^* is described by

$$I(\rho;\Lambda^*)\equiv\sup\left\{\sum_n S(\Lambda^*E_n, \Lambda^*\rho),\ \rho=\sum_n \lambda_n E_n\right\},$$

where $S(\cdot, \cdot)$ is the Umegaki's relative entropy and $\rho=\sum_n \lambda_n E_n$ represents a Schatten-von Neumann (one dimensional orthogonal) decomposition [49] of ρ. Since the Schatten-von Neumann decomposion of a state ρ is not unique unless all eigenvalues of ρ do not degenerate, the Ohya mutual entropy is defined by taking a supremum for all Schatten-von Neumann decomposion of a state ρ. Then the Ohya mutual entropy satisfies the following Shannon's type inequality [31]

$$0\le I(\rho, \Lambda^*)\le\min\{S(\rho), S(\Lambda^*\rho)\},$$

where $S(\rho)$ is the von Neumann entropy. This inequalities show that the Ohya mutual entropy represents the amount of information correctly carried from the input system to the output system through the quantum channel. The capacity denotes the ability of the information transmission of the communication processes, which was studied in [40,41,45].

For a certain set $\mathcal{S} \subset S(\mathcal{H}_1)$ satisfying some physical conditions, the capacity of quantum channel Λ^* [40] is defined by

$$C_q^{\mathcal{S}}(\Lambda^*) \equiv \sup\{I(\rho;\Lambda^*); \rho \in \mathcal{S}\}.$$

If $\mathcal{S} = S(\mathcal{H}_1)$ holds, then the capacity is denoted by $C_q(\Lambda^*)$. Then the following theorem for the attenuation channel was proved in [40].

Theorem For a subset $\mathcal{S}_n \equiv \{\rho \in S(\mathcal{H}_1); \dim s(\rho) = n\}$, the capacity of the attenuation channel Λ_0^* satisfies

$$C_q^{\mathcal{S}_n}(\Lambda_0^*) = \log n,$$

where $s(\rho)$ is the support projection of ρ.

10.6. Mutual entropy for general systems

Based on the classical relative entropy, the mutual entropy was discussed by Shannon to study the information transmission in classical systems and it was extended by Ohya [33,34,35] for fully general quantum systems.

Let $(\mathcal{A}, \mathfrak{S}(\mathcal{A}), \alpha(G))$ be a unital C^*-system and $\mathcal{S}$ be a weak* compact convex subset of $\mathfrak{S}(\mathcal{A})$. For an initial state $\varphi \in \mathcal{S}$ and a channel $\Lambda^* : \mathfrak{S}(\mathcal{A}) \to \mathfrak{S}(\mathcal{B})$, two compound states are

$$\Phi_\mu^{\mathcal{S}} = \int_{\mathcal{S}} \omega \otimes \Lambda^* \omega \, d\mu,$$

$$\Phi_0 = \varphi \otimes \Lambda^* \varphi.$$

The compound state $\Phi_\mu^{\mathcal{S}}$ expresses the correlation between the input state φ and the output state $\Lambda^*\varphi$. The mutual entropy with respect to $\mathcal{S}$ and μ is given by

$$I_\mu^{\mathcal{S}}(\varphi;\Lambda^*) = S(\Phi_\mu^{\mathcal{S}}, \Phi_0)$$

and the mutual entropy with respect to $\mathcal{S}$ is defined by Ohya [33] as

$$I^{\mathcal{S}}(\varphi;\Lambda^*) = \sup\{I_\mu^{\mathcal{S}}(\varphi;\Lambda^*); \mu \in M_\varphi(\mathcal{S})\}.$$

10.7. Mutual entropy type complexity

Shor [53] and Bennet et al [6,10] proposed the mutual type measures so-called the coherent entropy and the Lindblad-Nielson entropy by using the entropy exchange [50] defined by

$$S_e(\rho, \Lambda^*) = -trW\log W,$$

where W is a matrix $W = (W_{ij})_{i,j}$ with

$W_{ij} \equiv tr A_i^* \rho A_j$

for a state ρ concerning a Stinespring-Sudarshan-Kraus form

$\Lambda^*(\cdot) \equiv \sum_j A_j^* \cdot A_j,$

of a channel Λ^*. Then the coherent entropy $I_C(\rho; \Lambda^*)$ [53] and the Lindblad-Nielson entropy $I_L(\rho; \Lambda^*)$ [10] are given by

$I_C(\rho; \Lambda^*) \equiv S(\Lambda^* \rho) - S_e(\rho, \Lambda^*),$

$I_L(\rho; \Lambda^*) \equiv S(\rho) + S(\Lambda^* \rho) - S_e(\rho, \Lambda^*).$

In this section, we compare with these mutual types measures. By comparing these mutual entropies for quantum information communication processes, we have the following theorem [47]:

Theorem Let $\{A_j\}$ be a projection valued measure withdim $A_j = 1$. For arbitrary state ρ and the quantum channel $\Lambda^*(\cdot) \equiv \sum_j A_j \cdot A_j^*$, one has

1. $0 \le I(\rho; \Lambda^*) \le \min\{S(\rho), S(\Lambda^* \rho)\}$ (Ohya mutual entropy),
2. $I_C(\rho; \Lambda^*) = 0$ (coherent entropy),
3. $I_L(\rho; \Lambda^*) = S(\rho)$ (Lindblad-Nielsen entropy).

For the attenuation channel Λ_0^*, one can obtain the following theorems [47]:

Theorem For any state $\rho = \sum_n \lambda_n |n\rangle\langle n|$ and the attenuation channel Λ_0^* with $|\alpha|^2 = |\beta|^2 = \frac{1}{2}$, one has

1. $0 \le I(\rho; \Lambda_0^*) \le \min\{S(\rho), S(\Lambda_0^* \rho)\}$ (Ohya mutual entropy),
2. $I_C(\rho; \Lambda_0^*) = 0$ (coherent entropy),
3. $I_L(\rho; \Lambda_0^*) = S(\rho)$ (Lindblad-Nielsen entropy).

Theorem For the attenuation channel Λ_0^* and the input state $\rho = \lambda |0\rangle\langle 0| + (1-\lambda) |\theta\rangle\langle\theta|$, we have

1. $0 \le I(\rho; \Lambda_0^*) \le \min\{S(\rho), S(\Lambda_0^* \rho)\}$ (Ohya mutual entropy),
2. $-S(\rho) \le I_C(\rho; \Lambda_0^*) \le S(\rho)$ (coherent entropy),
3. $0 \le I_L(\rho; \Lambda_0^*) \le 2S(\rho)$ (Lindblad-Nielsen entropy)

The above Theorem shows that the coherent entropy $I_C(\rho;\Lambda_0^*)$ takes a minus value for $|\alpha|^2<|\beta|^2$ and the Lindblad-Nielsen entropy $I_L(\rho;\Lambda_0^*)$ is greater than the von Neumann entropy of the input state ρ for $|\alpha|^2>|\beta|^2$. Therefore Ohya mutual entropy is most suitable one for discussing the efficiency of information transmission in quantum processes. Since the above theorems and other results [47] we could conclude that Ohya mutual entropy might be most suitable one for discussing the efficiency of information transmission in quantum communication processes. It means that Ohya mutual entropy can be considered as the transmitted complexity for quantum communication processes.

11. Quantum dynamical entropy

The classical dynamical (or Kolmogorov-Sinai) entropy S(T) [23] for a measure preserving transformation T was defined on a message space through finite partitions of the measurable space.

The classical coding theorems of Shannon are important tools to analyse communication processes which have been formulated by the mean dynamical entropy and the mean dynamical mutual entropy. The mean dynamical entropy represents the amount of information per one letter of a signal sequence sent from an input source, and the mean dynamical mutual entropy does the amount of information per one letter of the signal received in an output system.

The quantum dynamical entropy (QDE) was studied by Connes-Størmer [13], Emch [15], Connes- Narnhofer-Thirring [12], Alicki-Fannes [3], and others [9,48,19,57,11]. Their dynamical entropies were defined in the observable spaces. Recently, the quantum dynamical entropy and the quantum dynamical mutual entropy were studied by the present authors [34,35]: (1) The dynamical entropy is defined in the state spaces through the complexity of Information Dynamics [36]. (2) It is defined through the quantum Markov chain (QMC) was done in [2]. (3) The dynamical entropy for a completely positive (CP) maps was introduced in [25].

12. Mean entropy and mean mutual entropy

The classical Shannon 's coding theorems are important subject to study communication processes which have been formulated by the mean entropy and the mean mutual entropy based on the classical dynamical entropy. The mean entropy shows the amount of information per one letter of a signal sequence of an input source, and the mean mutual entropy denotes the amount of information per one letter of the signal received in an output system. Those mean entropies were extended in general quantum systems.

In this section, we briefly explain a new formulation of quantum mean mutual entropy of K-S type given by Ohya [35,27].

In quantum information theory, a stationary information source is denoted by a C^* triple $(\mathcal{A}, \mathfrak{S}(\mathcal{A}), \theta_{\mathcal{A}})$ with a stationary state φ with respect to $\theta_{\mathcal{A}}$; that is, $\mathcal{A}$ is a unital C^*-algebra, $\mathfrak{S}(\mathcal{A})$ is the set of all states over $\mathcal{A}$, $\theta_{\mathcal{A}}$ is an automorphism of $\mathcal{A}$, and $\varphi \in \mathfrak{S}(\mathcal{A})$ is a state over $\mathcal{A}$ with $\varphi \circ \theta_{\mathcal{A}} = \varphi$.

Let an output C^*-dynamical system be the triple $(\mathcal{B}, \mathfrak{S}(\mathcal{B}), \theta_{\mathcal{B}})$, and $\Lambda^* : \mathfrak{S}(\mathcal{A}) \to \mathfrak{S}(\mathcal{B})$ be a covariant channel which is a dual of a completely positive unital map $\Lambda : \mathcal{B} \to \mathcal{A}$ such that $\Lambda \circ \theta_{\mathcal{B}} = \theta_{\mathcal{A}} \circ \Lambda$.

In this section, we explain functional $S_\mu^{\mathcal{S}}(\varphi; \alpha^M)$, $S^{\mathcal{S}}(\varphi; \alpha^M)$, $I_\mu^{\mathcal{S}}(\varphi; \alpha^M, \beta^N)$ and $I^{\mathcal{S}}(\varphi; \alpha^M, \beta^N)$ introduced in [35,27] for a pair of finite sequences of $\alpha^M = (\alpha_1, \alpha_2, \cdots, \alpha_M)$, $\beta^N = (\beta_1, \beta_2, \cdots, \beta_N)$ of completely positive unital maps $\alpha_m : \mathcal{A}_m \to \mathcal{A}$, $\beta_n : \mathcal{B}_n \to \mathcal{B}$ where $\mathcal{A}_m$ and $\mathcal{B}_n$ $(m=1, \cdots, M, n=1, \cdots, N)$ are finite dimensional unital C^*-algebras.

Let $\mathcal{S}$ be a weak * convex subset of $\mathfrak{S}(\mathcal{A})$ and φ be a state in $\mathcal{S}$. We denote the set of all regular Borel probability measures μ on the state space $\mathfrak{S}(\mathcal{A})$ of $\mathcal{A}$ by $M_\varphi(\mathcal{S})$, so that μ is maximal in the Choquet ordering and μ represents $\varphi = \int_{S(\mathcal{A})} \omega d\mu(\omega)$. Such measures is taken by extremal decomposition measures for φ, Using Choquet's theorem, one can be shown that there exits such measures for any state $\varphi \in \mathfrak{S}(\mathcal{A})$. For a given finite sequences of completely positive unital maps $\alpha_m : \mathcal{A}_m \to \mathcal{A}$ from finite dimensional unitalC^*-algebras $\mathcal{A}_m$ $(m=1, \cdots, M)$ and a given extremal decomposition measure μ of φ, the compound state of $\alpha_1^* \varphi, \alpha_2^* \varphi, \cdots, \alpha_M^* \varphi$ on the tensor product algebra $\bigotimes_{m=1}^{M} \mathcal{A}_m$ is given by [35,27]

$$\Phi_\mu^{\mathcal{S}}(\alpha^M) = \int_{S(\mathcal{A})} \bigotimes_{m=1}^{M} \alpha_m^* \omega d\mu(\omega).$$

Furthermore $\Phi_\mu^{\mathcal{S}}(\alpha^M \cup \beta^N)$ is a compound state of $\Phi_\mu^{\mathcal{S}}(\alpha^M)$ and $\Phi_\mu^{\mathcal{S}}(\beta^N)$ with $\alpha^M \cup \beta^N \equiv (\alpha_1, \alpha_2, \cdots, \alpha_M, \beta_1, \beta_2, \cdots, \beta_N)$ constructed as

$$\Phi_\mu^{\mathcal{S}}(\alpha^M \cup \beta^N) = \int_{S(A)} \left(\bigotimes_{m=1}^{M} \alpha_m^* \omega \right) \otimes \left(\bigotimes_{n=1}^{N} \beta_n^* \omega \right) d\mu$$

For any pair (α^M, β^N) of finite sequences $\alpha^M = (\alpha_1, \cdots, \alpha_M)$ and $\beta^N = (\beta_1, \cdots, \beta_N)$ of completely positive unital maps $\alpha_m : \mathcal{A}_m \to \mathcal{A}$, $\beta_n : \mathcal{B}_n \to \mathcal{A}$ from finite dimensional unital C^* -algebras and any extremal decomposition measure μ of φ, the entropy functional S_μ and the mutual entropy functional I_μ are defined in [35,27] such

$$S_\mu^S(\varphi;\alpha^M)=\int_{S(A)} S\left(\bigotimes_{m=1}^{M} \alpha_m^* \omega,\ \Phi_\mu^S(\alpha^M)\right) d\mu(\omega),$$

$$I_\mu^S(\varphi;\alpha^M, \beta^N)=S\left(\Phi_\mu^S(\alpha^M \cup \beta^N),\ \Phi_\mu^S(\alpha^M)\otimes\Phi_\mu^S(\beta^N)\right),$$

where $S(\cdot, \cdot)$ is the relative entropy.

For a given pair of finite sequences of completely positive unitalmaps $\alpha^M=(\alpha_1, \cdots, \alpha_M)$, $\beta^N=(\beta_1, \cdots, \beta_N)$, the functional $S^S(\varphi;\alpha^M)$ (resp. $I^S(\varphi;\alpha^M, \beta^N)$) is given in [35,27] by taking the supremum of $S_\mu^S(\varphi;\alpha^M)$ (resp. $I_\mu^S(\varphi;\alpha^M, \beta^N)$) for all possible extremal decompositions μ's of φ:

$$S^S(\varphi;\alpha^M)=\sup\left\{S_\mu^S(\varphi;\alpha^M);\quad \mu\in M_\varphi(\mathcal{S})\right\},$$

$$I^S(\varphi;\alpha^M, \beta^N)=\sup\left\{I_\mu^S(\varphi;\alpha^M, \beta^N);\quad \mu\in M_\varphi(\mathcal{S})\right\}.$$

Let $\mathcal{A}$ (resp. $\mathcal{B}$) be a unital C^*-algebra with a fixed automorphism $\theta_\mathcal{A}$ (resp. $\theta_\mathcal{B}$), Λ be a covariant completely positive unital map from $\mathcal{B}$ to $\mathcal{A}$, and φ be an invariant state over $\mathcal{A}$, i.e., $\varphi\circ\theta_\mathcal{A}=\varphi$.

$$\alpha^N\equiv\left(\alpha,\ \theta_\mathcal{A}\circ\alpha,\ \cdots,\ \theta_\mathcal{A}^{N-1}\circ\alpha\right),$$

$$\beta_\Lambda^N\equiv\left(\Lambda\circ\beta,\ \Lambda\circ\theta_\mathcal{B}\circ\beta,\ \cdots,\ \Lambda\circ\theta_\mathcal{B}^{N-1}\circ\beta\right).$$

For each completely positive unital map $\alpha : \mathcal{A}_0\to\mathcal{A}$ (resp. $\beta : \mathcal{B}_0\to\mathcal{B}$) from a finite dimensional unital C^*-algebra $\mathcal{A}_0$ (resp. $\mathcal{B}_0$) to $\mathcal{A}$ (resp. $\mathcal{B}$), $\tilde{S}^S(\varphi;\theta_\mathcal{A}, \alpha)$, $\tilde{I}^S(\varphi;\Lambda^*, \theta_\mathcal{A}, \theta_\mathcal{B}, \alpha, \beta)$ are given in [35,27] by

$$\tilde{S}^S(\varphi;\theta_\mathcal{A}, \alpha)=\liminf_{N\to\infty}\frac{1}{N}S^S(\varphi;\alpha^N),$$

$$\tilde{I}^S(\varphi;\Lambda^*, \theta_\mathcal{A}, \theta_\mathcal{B}, \alpha, \beta)=\liminf_{N\to\infty}\frac{1}{N}I^S(\varphi;\alpha^M, \beta^N).$$

The functional $\tilde{S}^S(\varphi;\theta_\mathcal{A})$ and $\tilde{I}^S(\varphi;\Lambda^*, \theta_\mathcal{A}, \theta_\mathcal{B})$ are defined by taking the supremum for all possible $\mathcal{A}_0$'s, α's, $\mathcal{B}_0$'s, and β's:

$$\tilde{S}^S(\varphi;\theta_\mathcal{A})=\sup_\alpha \tilde{S}^S(\varphi;\theta_\mathcal{A}, \alpha),$$

$$\tilde{I}^S(\varphi;\Lambda^*, \theta_\mathcal{A}, \theta_\mathcal{B})=\sup_{\alpha,\beta}\tilde{I}^S(\varphi;\Lambda^*, \theta_\mathcal{A}, \theta_\mathcal{B}, \alpha, \beta).$$

Then the fundamental inequality in information theory holds for $\tilde{S}^S(\varphi;\theta_\mathcal{A})$ and $\tilde{I}^S(\varphi;\Lambda^*, \theta_\mathcal{A}, \theta_\mathcal{B})$ [35].

12.1. Proposition

$$0\le\tilde{I}^S(\varphi;\Lambda^*, \theta_\mathcal{A}, \theta_\mathcal{B})\le\min\left\{\tilde{S}^S(\varphi;\theta_\mathcal{A}),\ \tilde{S}^S(\Lambda^*\varphi;\theta_\mathcal{B})\right\}.$$

These functional $\tilde{S}^{\mathcal{S}}(\varphi;\theta_{\mathcal{A}})$ and $\tilde{I}^{\mathcal{S}}(\varphi;\Lambda^*,\theta_{\mathcal{A}},\theta_{\mathcal{B}})$ are constructed from the functional $S_\mu^{\mathcal{S}}(\varphi;\alpha^N)$ and $I_\mu^{\mathcal{S}}(\varphi;\alpha^N,\beta^N)$ coming from information theory and these functionals are obtained by using a channel transformation, so that those functionals contains the dynamical entropy as a special case [35,27]. Moreover these functionals contain usual K-S entropies as follows [35,27].

Proposition If $\mathcal{A}_k$, $\mathcal{A}$ are abelian C^*-algebras and each α_k is an embedding, then our functionals coincide with classical K-S entropies:

$$S_\mu^{\mathfrak{S}(\mathcal{A})}(\varphi;\alpha^M)=S_\mu^{classical}\left(\bigvee_{m=1}^{M}\tilde{A}_m\right),$$

$$I_\mu^{\mathfrak{S}(\mathcal{A})}(\varphi;\alpha^M,\beta_{id}^N)=I_\mu^{classical}\left(\bigvee_{m=1}^{M}\tilde{A}_m,\bigvee_{n=1}^{N}\tilde{B}_n\right)$$

for any finite partitions $\tilde{A}_m$, $\tilde{B}_n$ of a probability space $(\Omega,\mathfrak{F},\varphi)$.

In general quantum structure, we have the following theorems [35,27].

Theorem Let α_m be a sequence of completely positive maps $\alpha_m:\mathcal{A}_m\to\mathcal{A}$ such that there exist completely positive maps $\alpha_m':\mathcal{A}\to\mathcal{A}_m$ satisfying $\alpha_m\circ\alpha_m'\to id_{\mathcal{A}}$ in the pointwise topology. Then:

$$\tilde{S}^{\mathcal{S}}(\varphi;\theta_{\mathcal{A}})=\lim_{m\to\infty}\tilde{S}^{\mathcal{S}}(\varphi;\theta_{\mathcal{A}},\alpha_m).$$

Theorem Let α_m and β_m be sequences of completely positive maps $\alpha_m:\mathcal{A}_m\to\mathcal{A}$ and $\beta_m:\mathcal{B}_m\to\mathcal{B}$ such that there exist completely positive maps $\alpha_m':\mathcal{A}\to\mathcal{A}_m$ and $\beta_m':\mathcal{B}\to\mathcal{B}_m$ satisfying $\alpha_m\circ\alpha_m'\to id_{\mathcal{A}}$ and $\beta_m\circ\beta_m'\to id_{\mathcal{B}}$ in the pointwise topology. Then one has

$$\tilde{I}^{\mathcal{S}}(\varphi;\Lambda^*,\theta_{\mathcal{A}},\theta_{\mathcal{B}})=\lim_{m\to\infty}\tilde{I}^{\mathcal{S}}(\varphi;\Lambda^*,\theta_{\mathcal{A}},\theta_{\mathcal{B}},\alpha_m,\beta_m).$$

The above theorem is a Kolmogorov-Sinai type convergence theorem for the mutual entropy [35,27,28,34].

In particular, a quantum extension of classical formulation for information transmission giving a basis of Shannon's coding theorems can be considered in the case that $A=\bigotimes_{-\infty}^{\infty}A_0$, $B=\bigotimes_{-\infty}^{\infty}B_0$, $S=\mathfrak{S}$ and θ_A, θ_B are shift operators, both denoted by θ. In this case, the channel capacity is defined as [40,41,45,46,38,39,42,43]

$$\tilde{C}(\Lambda^*)\equiv\sup\{\tilde{I}^{\mathfrak{S}}(\varphi;\Lambda^*,\theta);\varphi\in\mathfrak{S}\}.$$

Using this capacity, one can consider Shannon's coding theorems in fully quantum systems.

13. Computations of mean entropies for modulated states

Based on the paper [59], we here explain general modulated states and briefly review some examples of modulated states (PPM, OOK, PSK).

Let $\{a_1, \cdots, a_N\}$ be an alphabet set constructing the input signals and $N \equiv \{E_1, \cdots, E_N\}$ be the set of one dimensional projections on a Hilbert space $\mathcal{H}$ satisfying

1. $E_n \perp E_m \quad (n \neq m)$
2. E_n corresponds to the alphabet a_n.

We denote the set of all density operators on $\mathcal{H}$ generated by

$$\mathfrak{S}_0 \equiv \left\{\rho_0 = \sum_{n=1}^{N} \lambda_n E_n; \rho_0 \geq 0,\ tr\rho_0 = 1\right\},$$

where an element of $\mathfrak{S}_0$ represents a state of the quantum input system. The state is transmitted from the quantum input system to the quantum modulator in order to send information effectively, whose transmitted state is called the quantum modulated state. The quantum modulated states are denoted as follows: Let M be an ideal modulator and $N \equiv \{E_1^{(M)}, \cdots, E_N^{(M)}\}$ be the set of one dimensional projections on a Hilbert space $\mathcal{H}_M$ for modulated signals satisfying

$E_n^{(M)} \perp E_m^{(M)} (n \neq m)$, and we represent the set of all density operators on $\mathcal{H}_M$ by

$$\mathfrak{S}_0^{(M)} \equiv \left\{\rho_0^{(M)} = \sum_{n=1}^{N} \mu_n E_n^{(M)}; \quad \rho_0^{(M)} \geq 0, \quad tr\rho_0^{(M)} = 1\right\},$$

where an element of $\mathfrak{S}_0^{(M)}$ represents a modulated state of the quantum input system. There are many expressions for the modulations. In this section, we take the modulated states by means of the photon number states.

γ_M^* is a modulator M if $\gamma_M^*(E_n) = E_n^{(M)}$ is a map from $\mathfrak{S}_0$ to $\mathfrak{S}_0^{(M)}$ satisfying (1) γ_M is a completely positive unital map from $\mathcal{A}_0$ to $\mathcal{A}$. Moreover γ_{IM}^* is called an ideal modulator IM if (1) $\gamma_{IM}^*(E_n) = E_n^{(M)}$ is a modulator from $\mathfrak{S}_0$ to $\mathfrak{S}_0^{(M)}$, $\gamma_{IM}^*(E_n) \perp \gamma_{IM}^*(E_m)$ for any orthogonal $E_n \in \mathfrak{S}_0$. Some examples of ideal modulator are given as follows:

1. For any $E_n \in \mathfrak{S}_0$, the PPM (Pulse Position Modulator) is defined by

$$\gamma_{PPM}^*(E_n) \equiv E_n^{(PPM)}$$
$$= E_0^{PAM} \otimes \cdots \otimes E_0^{(PAM)} \otimes E_d^{(PAM)} \otimes E_0^{(PAM)} \otimes \cdots E_0^{(PAM)}$$

where $E_0^{(PAM)}$ is the vacuum state on $\mathcal{H}_{(PAM)}$.

2. For $E_1, E_2 \in \mathfrak{S}_0$, the OOK (On-Off Keying) is defined by

$\gamma_{OOK}^{*}(E_1) \equiv E_1^{(OOK)} = |0><0|,$

$\gamma_{OOK}^{*}(E_2) \equiv E_2^{(OOK)} = |\kappa><\kappa|$

where $|\kappa><\kappa|$ is the coherent state on $\mathcal{H}_{OOK}$.

3. For $E_1, E_2 \in S_0$, the PSK (Phase Shift Keying) is defined by

$\gamma_{PSK}^{*}(E_1) \equiv E_1^{(PSK)} = |-\kappa><-\kappa|,$

$\gamma_{PSK}^{*}(E_2) \equiv E_2^{(PSK)} = |\kappa><\kappa|$

where $|\kappa><\kappa|$, $|-\kappa><-\kappa|$ are the coherent states on $\mathfrak{H}_{PSK}$.

Now we briefly review the calculation of the mean mutual entropy of K-S type for the modulated state (PSK) by means of the coherent state. Other calculations are obtained in [59].

$\alpha_{(IM)}^{N}, \beta_{(IM)}^{N}$ are given by

$\alpha_{(IM)}^{N} \equiv (\alpha \circ \tilde{\gamma}_{(IM)}, \theta_{\mathcal{A}} \circ \alpha \circ \tilde{\gamma}_{(IM)}, \cdots, \theta_{\mathcal{A}}^{N-1} \circ \alpha \circ \tilde{\gamma}_{(IM)}),$

$\beta_{(IM)}^{N} \equiv (\tilde{\gamma}_{(IM)} \circ \Lambda \circ \beta, \tilde{\gamma}_{(IM)} \circ \Lambda \circ \theta_{\mathcal{B}} \circ \beta, \cdots, \tilde{\gamma}_{(IM)} \circ \Lambda \circ \theta_{\mathcal{B}}^{N-1} \circ \beta),$

where $\tilde{\Lambda} \equiv \bigotimes_{i=-\infty}^{\infty} \Lambda$ and $\tilde{\gamma}_{(IM)} \equiv \bigotimes_{i=-\infty}^{\infty} \gamma_{(IM)}$ are held.

PSK. For an initial state $\rho = \bigotimes_{i=-\infty}^{\infty} \rho_i \in \bigotimes_{i=-\infty}^{\infty} S_i$, let $\rho_i = \nu|-\kappa\rangle\langle-\kappa| + (1-\nu)|\kappa\rangle\langle\kappa|$ $(0 \le \nu \le 1)$. The Schatten decomposition of ρ_i is obtained as

$$\rho_i = \sum_{n_i=1}^{2} \lambda_{n_i} E_{n_i}^{(PSK)},$$

where the eigenvalues λ_{n_i} of ρ_i are

$$\lambda_1 = \frac{1}{2}\left\{1 + \sqrt{1 - 4\nu(1-\nu)\left(1 - \exp(-|2\kappa|^2)\right)}\right\},$$

$$\lambda_2 = \frac{1}{2}\left\{1 - \sqrt{1 - 4\nu(1-\nu)\left(1 - \exp(-|2\kappa|^2)\right)}\right\}.$$

Two projections $E_{n_i}^{(PSK)}$ $(n_i = 1, 2)$ and the eigenvectors $|e_{n_i}^{(PSK)}\rangle$ of λ_{n_i} $(n_i = 1, 2)$ are given by

$$E_{n_i}^{(PSK)} = |e_{n_i}^{(PSK)}\rangle\langle e_{n_i}^{(PSK)}|,$$

$$|e_{n_i}^{(PSK)}\rangle = a_{n_i}|-\kappa\rangle + b_{n_i}|\kappa\rangle, \quad (n_i = 1, 2),$$

where

$$|b_{n_i}|^2 = \frac{1}{\tau_{n_i}^2 + 2\exp(-|\kappa|^2)\tau_{n_i} + 1},$$

$$|a_{n_i}|^2 = \tau_{n_i}^2 |b_{n_i}|^2,$$

$$a_{n_i}\overline{b_{n_i}} = \overline{a_{n_i}} b_{n_i} = \tau_{n_i} |b_{n_i}|^2$$

$$\tau_1 = \frac{-(1-2\nu) + \sqrt{1-4\nu(1-\nu)(1-\exp(-|2\kappa|^2))}}{2(1-\nu)\exp(-|\kappa|^2)},$$

$$\tau_2 = \frac{-(1-2\nu) - \sqrt{1-4\nu(1-\nu)(1-\exp(-|2\kappa|^2))}}{2(1-\nu)\exp(-|\kappa|^2)}.$$

For the above initial state $E_{n_i}^{(PSK)}$, one can obtain the output state for the attenuation channel Λ^* as follows:

$$\Lambda^* E_{n_i}^{(PSK)} = \sum_{n_i'=1}^{2} \hat{\lambda}_{n_i, n_i'} \hat{E}_{n_i, n_i'}^{(PSK)} \quad (n_i = 1,\ 2),$$

where the eigenvalues $\hat{\lambda}_{n_i, n_i'}$ of $\Lambda^* E_{n_i}^{(OOK)}$ are given by ($n_i = 1,\ 2$)

$$\hat{\lambda}_{n_i, 1} = \frac{1}{2}\left\{1 + \sqrt{1-4\mu_{n_i}(1-\mu_{n_i})(1-|\langle u_{n_i, 1}, u_{n_i, 2}\rangle|^2)}\right\},$$

$$\hat{\lambda}_{n_i, 2} = \frac{1}{2}\left\{1 - \sqrt{1-4\mu_{n_i}(1-\mu_{n_i})(1-|\langle u_{n_i, 1}, u_{n_i, 2}\rangle|^2)}\right\},$$

$$\mu_{n_i} = \frac{1}{2}(1+\exp(-(1-\eta)|\kappa|^2)) \frac{\tau_{n_i}^2 + 2\exp(-|\alpha\kappa|^2)\tau_{n_i} + 1}{\tau_{n_i}^2 + 2\exp(-|\kappa|^2)\tau_{n_i} + 1}.$$

$$|\langle u_{n_i, n_i'}, u_{n_i, n_i'}\rangle|^2 = 1,$$

$$\langle u_{n_i, 1}, u_{n_i, 2}\rangle = \frac{\tau_{n_i}^2 - 1}{\sqrt{(\tau_{n_i}^2+1)^2 - 4\exp(-|2\alpha\kappa|^2)\tau_{n_i}^2}} \quad (n_i = 1,\ 2).$$

$\hat{E}_{n_i, n_i'}^{(PSK)}$ are the eigenstates with respect to $\hat{\lambda}_{n_i, n_i'}$. Then we have

$$\Phi_E(\alpha_{(PSK)}^N) = \bigotimes_{i=0}^{N-1} \gamma_{(PSK)}^* \circ \alpha^* \circ \theta_A^{*i}(\rho) = \bigotimes_{i=0}^{N-1} \gamma_{(PSK)}^*(\rho_i)$$

$$= \sum_{n_0=1}^{2} \cdots \sum_{n_{N-1}=1}^{2} \left(\prod_{k=0}^{N-1} \lambda_{n_k}\right)\left(\bigotimes_{i=0}^{N-1} E_{n_i}^{(PSK)}\right)$$

When Λ^* is given by the attenuation channel, we get

$$\Phi_E(\beta^N_{\Lambda(PSK)}) = \bigotimes_{i=0}^{N-1} \beta^* \circ \theta_B^{*i} \circ \Lambda^* \circ \gamma^*_{(PSK)}(\rho)$$

$$= \sum_{n_0=1}^{2} \cdots \sum_{n_{N-1}=1}^{2} \left(\prod_{k=0}^{N-1} \lambda_{n_k} \right) \left(\bigotimes_{i=0}^{N-1} \Lambda^* E_{n_i}^{(PSK)} \right)$$

The compound states through the attenuation channel Λ^* becomes

$$\Phi_E(\alpha^N_{(PSK)} \cup \beta^N_{\Lambda(PSK)})$$

$$= \sum_{n_0=1}^{2} \cdots \sum_{n_{N-1}=1}^{2} \left(\prod_{k=0}^{N-1} \lambda_{n_k} \right) \sum_{n'_0=1}^{2} \cdots \sum_{n'_{N-1}=1}^{2} \left(\prod_{k'=0}^{N-1} \hat{\lambda}_{n_{k'}, n'_{k'}} \right)$$

$$\times \left(\bigotimes_{i=0}^{N-1} E_{n_i}^{(PSK)} \right) \otimes \left(\bigotimes_{i'=0}^{N-1} \hat{E}^{(PSK)}_{n_{i'}, n'_{i'}} \right)$$

$$\Phi_E(\alpha^N_{(PSK)}) \otimes \Phi_E(\beta^N_{\Lambda(PSK)})$$

$$= \sum_{n_0=1}^{2} \cdots \sum_{n_{N-1}=1}^{2} \left(\prod_{k=0}^{N-1} \lambda_{n_k} \right) \sum_{m_0=1}^{2} \cdots \sum_{m_{N-1}=1}^{2} \left(\prod_{k'=0}^{N-1} \lambda_{m_{k'}} \right)$$

$$\times \sum_{m'_0=0}^{2} \cdots \sum_{m'_{N-1}=0}^{2} \left(\prod_{k''=0}^{N-1} \hat{\lambda}_{m_{k''}, m'_{k''}} \right)$$

$$\times \left(\bigotimes_{i=0}^{N-1} E_{n_i}^{(PSK)} \right) \otimes \left(\bigotimes_{i'=0}^{N-1} \hat{E}^{(PSK)}_{m_{i'}, m'_{i'}} \right)$$

Lemma For an initial state $\rho = \bigotimes_{i=-\infty}^{\infty} \rho_i \in \bigotimes_{i=-\infty}^{\infty} S_i$, we have

$$I_E(\rho; \alpha^N_{(PSK)}, \beta^N_{(PSK)})$$

$$= \sum_{n'_0=1}^{2} \cdots \sum_{n'_{N-1}=1}^{2} \sum_{n_0=1}^{2} \cdots \sum_{n_{N-1}=1}^{2} \left(\prod_{k=0}^{N-1} \lambda_{n_k} \hat{\lambda}_{n_k, n'_k} \right) \log \frac{\prod_{k=0}^{N-1} \hat{\lambda}_{n_k, n'_k}}{\sum_{m_0=1}^{2} \cdots \sum_{m_{N-1}=1}^{2} \left(\prod_{k'=0}^{N-1} \lambda_{m_{k'}} \hat{\lambda}_{m_{k'}, n'_{k'}} \right)}.$$

By using the above lemma, we have the following theorem.

Theorem For an initial state $\rho = \bigotimes_{i=-\infty}^{\infty} \rho_i \in \bigotimes_{i=-\infty}^{\infty} S_i$, we have

$$\tilde{S}(\rho; \theta_{\Lambda}, \alpha^N_{(PSK)}) = \lim_{N \to \infty} \frac{1}{N} S(\rho; \alpha^N_{(PSK)}) = -\sum_{n=1}^{2} \lambda_n \log \lambda_n$$

and

$$\tilde{I}(\rho;\Lambda^*,\theta_{\curlyvee},\theta_{\curlyvee},\alpha_{(PSK)}^N,\beta_{(PSK)}^N)=\sum_{n'=1}^{2}\sum_{n=1}^{2}\lambda_n\hat{\lambda}_{n,n'}\log\frac{\hat{\lambda}_{n,n'}}{\sum_{m=1}^{2}\lambda_m\hat{\lambda}_{m,n'}}.$$

14. KOW dynamical entropy

In this section, we briefly explain the definition of the KOW entropy according to [25].

For a normal state ω on $B(\mathcal{K})$ and a normal, unital CP linear map Γ from $B(\mathcal{K})\otimes B(\mathcal{H})$ to $B(\mathcal{K})\otimes B(\mathcal{H})$, one can define a transition expectation $E^{\Gamma,\omega}$ from $B(\mathcal{K})\otimes B(\mathcal{H})$ to $B(\mathcal{H})$ by

$$E^{\Gamma,\omega}(\widetilde{A})=\omega(\Gamma(\widetilde{A}))=tr_{\mathcal{K}}\widetilde{\omega}\Gamma(\widetilde{A}),\quad\forall\,\widetilde{A}\in B(\mathcal{K})\otimes B(\mathcal{H})$$

in the sense of [1,25], where $\widetilde{\omega}\in\mathfrak{S}(\mathcal{K})$ is a density operator associated to ω. The dual map E is a lifting from $\mathfrak{S}(\mathcal{H})$ to $\mathfrak{S}(\mathcal{K}\otimes\mathcal{H})$ by

$$E^{*\Gamma,\omega}(\rho)=\Gamma^*(\widetilde{\omega}\otimes\rho).$$

in the sense of Accardi and Ohya [1]. For a normal, unital CP map Λ from $B(\mathcal{H})$ to $B(\mathcal{H})$ and the identity map id on $B(\mathcal{K})$, the transition expectation

$$E_{\Lambda}^{\Gamma,\omega}(\widetilde{A})=\omega((id\otimes\Lambda)\Gamma(\widetilde{A})),\quad\forall\,\widetilde{A}\in B(\mathcal{K})\otimes B(\mathcal{H})$$

and the lifting is defined by

$$E_{\Lambda}^{*\Gamma,\omega}(\rho)=\Gamma^*(\widetilde{\omega}\otimes\Lambda^*(\rho)),\quad\forall\,\rho\in\mathfrak{S}(\mathcal{H}),$$

where $id\otimes\Lambda$ is a normal, unital CP map from $B(\mathcal{K})\otimes B(\mathcal{H})$ to $B(\mathcal{K})\otimes B(\mathcal{H})$ and Λ^* is a quantum channel [30,21,31,39,44,46,43] from $\mathfrak{S}(\mathcal{H})$ to $\mathfrak{S}(\mathcal{H})$ with respect to an input signal state ρ and a noise state $\widetilde{\omega}$. Based on the following relation

$$\begin{aligned}&tr_{(\otimes_1^n\mathcal{K})\otimes\mathcal{H}}\Phi_{\Lambda,n}^{*\Gamma,\omega}(\rho)(A_1\otimes\cdots\otimes A_n\otimes B)\\&\equiv tr_{\mathcal{H}}\rho\left(E_{\Lambda}^{\Gamma,\omega}\left(A_1\otimes E_{\Lambda}^{\Gamma,\omega}\left(A_2\otimes\cdots A_{n-1}\otimes E_{\Lambda}^{\Gamma,\omega}(A_n\otimes B)\cdots\right)\right)\right)\end{aligned}$$

for all $A_1,A_2,\cdots,A_n\in B(\mathcal{K})$, $B\in B(\mathcal{H})$ and any $\rho\in\mathfrak{S}(\mathcal{H})$, a lifting $\Phi_{\Lambda,n}^{*\Gamma,\omega}$ from $\mathfrak{S}(\mathcal{H})$ to $\mathfrak{S}\left((\otimes_1^n\mathcal{K})\otimes\mathcal{H}\right)$ and marginal states are given by

$$\rho_{\Lambda,n}^{\Gamma,\omega}\equiv tr_{\mathcal{H}}\Phi_{\Lambda,n}^{*\Gamma,\omega}(\rho)\in\mathfrak{S}\left(\otimes_1^n\mathcal{K}\right)\text{ and }\bar{\rho}_{\Lambda,n}^{\Gamma,\omega}\equiv tr_{\otimes_1^n\mathcal{K}}\Phi_{\Lambda,n}^{*\Gamma,\omega}(\rho)\in\mathfrak{S}(\mathcal{H})$$

where $\Phi_{\Lambda,n}^{*\Gamma,\omega}(\rho)$ is a compound state with respect to $\bar{\rho}_{\Lambda,n}^{\Gamma,\omega}$ and $\rho_{\Lambda,n}^{\Gamma,\omega}$ in the sense of [25,31].

Definition The quantum dynamical entropy with respect to Λ, ρ, Γ and ω is defined by

$$\tilde{S}(\Lambda;\rho,\Gamma,\omega)\equiv\limsup_{n\to\infty}\frac{1}{n}S\left(\rho_{\Lambda,n}^{\Gamma,\omega}\right),$$

where $S(\rho_{\Lambda,n}^{\Gamma,\omega})$ is the von Neumann entropy of $\rho_{\Lambda,n}^{\Gamma,\omega} \in \mathfrak{S}(\otimes_1^n \mathcal{K})$ defined by $S(\rho_{\Lambda,n}^{\Gamma,\omega}) = -tr\rho_{\Lambda,n}^{\Gamma,\omega}\log\rho_{\Lambda,n}^{\Gamma,\omega}$. The dynamical entropy with respect to Λ and ρ is defined as $\tilde{S}(\Lambda;\rho) \equiv \sup\{\mathfrak{S}(\Lambda;\rho,\Gamma,\omega);\Gamma,\omega\}$.

15. Formulation of generalized AF and AOW entropies by KOW entropy

In this section, I briefly explain the generalized AF and AOW entropies based on the KOW entropy [25].

For a finite operational partition of unity $\gamma_1, \cdots, \gamma_d \in B(\mathcal{H})$, i.e., $\sum_{i=1}^{d}\gamma_i^*\gamma_i = I$, and a normal unital CP map Λ from $B(\mathcal{H})$ to $B(\mathcal{H})$, transition expectations E_Λ^γ from $M_d \otimes B(\mathcal{H})$ to $B(\mathcal{H})$ and $E_\Lambda^{\gamma(0)}$ from $M_d^0 \otimes B(\mathcal{H})$ to $B(\mathcal{H})$ are defined by

$$E_\Lambda^\gamma\left(\sum_{i,j=1}^{d} E_{ij} \otimes A_{ij}\right) \equiv \sum_{i,j=1}^{d} \Lambda(\gamma_i^* A_{ij}\gamma_j),$$

$$E_\Lambda^{\gamma(0)}\left(\sum_{i,j=1}^{d} E_{ij} \otimes A_{ij}\right) \equiv \sum_{i=1}^{d} \Lambda(\gamma_i^* A_{ii}\gamma_i),$$

where $E_{ij} = |e_i\rangle\langle e_j|$ with normalized vectors $e_i \in \mathcal{H}$, $i = 1, 2, \cdots, d \le \dim\mathcal{H}$, M_d in $B(\mathcal{H})$ is the $d \times d$ matrix algebra and M_d^0 is a subalgebra of M_d consisting of diagonal elements of M_d. Then the quantum Markov states

$$\rho_{\Lambda,n}^\gamma = \sum_{i_1,\cdots,i_n=1}^{d} \sum_{j_1,\cdots,j_n=1}^{d} tr_{\mathcal{H}}\rho\Lambda(W_{j_1i_1}(\Lambda(W_{j_2i_2}(\cdots\Lambda(W_{j_ni_n}(I_{\mathcal{H}}))))))$$
$$\times E_{i_1j_1} \otimes \cdots \otimes E_{i_nj_n}$$

and $\rho_{\Lambda,n}^{\gamma(0)}$ is obtained by

$$\rho_{\Lambda,n}^{\gamma(0)} = \sum_{i_1,\cdots,i_n=1}^{d} tr_{\mathcal{H}}\rho\Lambda(W_{i_1i_1}(\Lambda(W_{i_2i_2}(\cdots\Lambda(W_{i_ni_n}(I_{\mathcal{H}}))))))E_{i_1i_1} \otimes \cdots \otimes E_{i_ni_n}$$
$$= \sum_{i_1,\cdots,i_n=1}^{d} p_{i_1,\cdots,i_n} E_{i_1i_1} \otimes \cdots \otimes E_{i_ni_n},$$

where

$$W_{ij}(A) \equiv \gamma_i^* A\gamma_j, \quad A \in B(\mathcal{H}),$$
$$W_{ij}^*(\rho) \equiv \gamma_j\rho\gamma_i^*, \quad \rho \in \mathfrak{S}(\mathcal{H}),$$
$$p_{i_1,\cdots,i_n} \equiv tr_{\mathcal{H}}\rho\Lambda(W_{i_1i_1}(\Lambda(W_{i_2i_2}(\cdots\Lambda(W_{i_ni_n}(I_{\mathcal{H}}))))))$$
$$= tr_{\mathcal{H}}W_{i_ni_n}^*(\Lambda^*\cdots\Lambda^*(W_{i_2i_2}^*(\Lambda^*(W_{i_1i_1}^*(\Lambda^*(\rho)))))).$$

Therefore the generalized AF entropy $\tilde{S}_{\mathcal{B}}(\Lambda;\rho)$ and the generalized AOW entropy $\tilde{S}_{\mathcal{B}}^{(0)}(\Lambda;\rho)$ of Λ and ρ with respect to a finite dimensional subalgebra $\mathcal{B}\subset B(\mathcal{H})$ are obtained by

$$\tilde{S}_{\mathcal{B}}(\Lambda;\rho)\equiv\sup_{\{\gamma_i\}\subset\mathcal{B}}\tilde{S}(\Lambda;\rho,\{\gamma_i\}),$$

$$\tilde{S}_{\mathcal{B}}^{(0)}(\Lambda;\rho)\equiv\sup_{\{\gamma_i\}\subset\mathcal{B}}^{(0)}\tilde{S}(\Lambda;\rho,\{\gamma_i\}),$$

where the dynamical entropies $\tilde{S}(\Lambda;\rho,\{\gamma_i\})$ and $\tilde{S}^{(0)}(\Lambda;\rho,\{\gamma_i\})$ are given by

$$\tilde{S}(\Lambda;\rho,\{\gamma_i\})\equiv\limsup_{n\to\infty}\frac{1}{n}S(\rho_{\Lambda,n}^{\gamma}),$$

$$\tilde{S}^{(0)}(\Lambda;\rho,\{\gamma_i\})\equiv\limsup_{n\to\infty}\frac{1}{n}S(\rho_{\Lambda,n}^{\gamma(0)}).$$

16. Computations of generalized AOW entropy for modulated states

Then we have the following theorem [25]:

16.1. Theorem

$$\tilde{S}_{\mathcal{B}}(\Lambda;\rho)\leq\tilde{S}_{\mathcal{B}}^{(0)}(\Lambda;\rho).$$

$\tilde{S}_{\mathcal{B}}^{(0)}(\Lambda;\rho)$ is equal to the AOW entropy if $\{\gamma_i\}$ is PVM (projection valued measure) and Λ is given by an automorphism θ. $\tilde{S}_{\mathcal{B}}(\Lambda;\rho)$ is equal to the AF entropy if $\{\gamma_i{}^*\gamma_i\}$ is POV (positive operator valued measure) and Λ is given by an automorphism θ. For the noisy optical channel, the generalized AOW entropy can be obtained in [58] as follows.

Theorem [58] When ρ is given by $\rho=\lambda|0\rangle\langle 0|+(1-\lambda)|\xi\rangle\langle\xi|$ and Λ^* is the noisy optical channel with the cohetent noise $|\kappa\rangle\langle\kappa|$ and parameters α, β satisfying $|\alpha|^2+|\beta|^2=1$, the quantum dynamical entropy with respect to Λ, ρ and $\{\gamma_j\}$ is obtained by

$$\tilde{S}^{(0)}(\Lambda;\rho,\{\gamma_j\})=-\sum_{j,k}q_{k,j}q_j\log q_{k,j},$$

where

$$q_j=\lambda|\langle\beta\kappa,x_j\rangle|^2+(1-\lambda)|\langle\alpha\xi+\beta\kappa,x_j\rangle|^2$$

$$q_{k,j}=v_j^+|\langle x_k,y_j^+\rangle|^2+(1-v_j^+)|\langle x_k,y_j^-\rangle|^2,$$

$$|y_j^+\rangle=a_j^+|\beta\kappa\rangle+b_j^+|\alpha\xi+\beta\kappa\rangle,$$

$$|y_j^-\rangle=a_j^-|\beta\kappa\rangle-b_j^-|\alpha\xi+\beta\kappa\rangle,$$

$$a_j^+ = \varepsilon_j^+ a_j,$$
$$a_j^- = \varepsilon_j^- a_j,\ b_j^+ = \varepsilon_j^+ b_j,$$
$$b_j^- = \varepsilon_j^- b_j,$$
$$\varepsilon_j^+ = \sqrt{\frac{\tau_j^2 + 2\exp\left(-\frac{1}{2}|\xi|^2\right)\tau_j + 1}{\tau_j^2 + 2\exp\left(-\frac{1}{2}|\alpha\xi|^2\right)\tau_j + 1}},$$
$$\varepsilon_j^- = \sqrt{\frac{\tau_j^2 + 2\exp\left(-\frac{1}{2}|\xi|^2\right)\tau_j + 1}{\tau_j^2 - 2\exp\left(-\frac{1}{2}|\alpha\xi|^2\right)\tau_j + 1}},$$
$$v_j^+ = \frac{1}{2}\left(1 + \exp\left(-\frac{1}{2}(1-|\alpha|^2)|\xi|^2\right)\right)\frac{1}{(\varepsilon_j^+)^2},$$

$$\tau_j = \frac{-(1-2\lambda)}{2(1-\lambda)\exp\left(-\frac{1}{2}|\xi|^2\right)} + (-1)^j \frac{\sqrt{1-4\lambda(1-\lambda)(1-\exp(-|\xi|^2))}}{2(1-\lambda)\exp\left(-\frac{1}{2}|\xi|^2\right)},$$

$$|b_j|^2 = \frac{1}{\tau_j^2 + 2\exp\left(-\frac{1}{2}|\xi|^2\right)\tau_j + 1},$$
$$|a_j|^2 = \tau_j^2 |b_j|^2,\ \bar{a}_j b_j = a_j \bar{b}_j = \tau_j |b_j|^2$$

Theorem [58] For $n \geqq 3$, the above compound state $\rho_{\Lambda,n}^{\gamma(0)}$ is written by

$$\rho_{\Lambda,n}^{\gamma(0)} = \sum_{j_1,\cdots,j_n=1}^{2} q_{j_1,\cdots,j_n} \bigotimes_{k=1}^{n} |x_{j_k}\rangle\langle x_{j_k}|,$$

where

$$q_{j_1,\cdots,j_n} \equiv tr_{\mathcal{H}} W_{j_n j_n}^*\left(\Lambda^*\left(\cdots\Lambda^*\left(W_{j_2 j_2}^*\left(\Lambda^*\left(W_{j_1 j_1}^*(\Lambda^*(\rho))\right)\right)\right)\cdots\right)\right),$$
$$\Lambda^*(\rho) = \lambda|\beta\kappa\rangle\langle\beta\kappa| + (1-\lambda)|\alpha\xi+\beta\kappa\rangle\langle\alpha\xi+\beta\kappa|,$$
$$W_{jj}^*(\Lambda^*(\rho)) \equiv \gamma_j^* \Lambda^*(\rho)\gamma_j = \left(\lambda|\langle\beta\kappa, x_j\rangle|^2 + (1-\lambda)|\langle\alpha\xi+\beta\kappa, x_j\rangle|^2\right)|x_j\rangle\langle x_j|.$$

Based on [40,41,45,46], one can obtain

$$\Lambda^*(|x_j\rangle\langle x_j|) = v_j^+ |y_j^+\rangle\langle y_j^+| + (1-v_j^+)|y_j^-\rangle\langle y_j^-|.$$

Thus we have

$$q_{j_1,\cdots,j_n} = \prod_{k=2}^{n}\left(v_{j_{k-1}}^+ |\langle x_{j_k}, y_{j_{k-1}}^+\rangle|^2 + \left(1-v_{j_{k-1}}^+\right)|\langle x_{j_k}, y_{j_{k-1}}^-\rangle|^2\right)$$
$$\times\left(\lambda|\langle\beta\kappa, x_j\rangle|^2 + (1-\lambda)|\langle\alpha\xi+\beta\kappa, x_j\rangle|^2\right)$$
$$= \prod_{k=2}^{n} q_{j_k, j_{k-1}} q_{j_1}.$$

If ${}_j q_{k,j} q_j = q_k$ is hold , then we get the dynamical entropy with respect to Λ, ρ and $\{\gamma_j\}$ such as

$$\tilde{S}^{(0)}(\Lambda;\rho,\{\gamma_j\}) = -\sum_{j,k} q_{k,j} q_j \log q_{k,j}.$$

Author details

Noboru Watanabe*

Address all correspondence to: watanabe@is.noda.tus.ac.jp

Department of Information Sciences, Tokyo University of Science, Noda City, Chiba, Japan

References

[1] Accardi, L., and Ohya, M., Compondchannnels, transition expectation and liftings, Appl. Math, Optim., 39, 33-59 (1999).

[2] Accardi, L., Ohya, M., and Watanabe, N., Dynamical entropy through quantum Markov chain, Open System and Information Dynamics, 4, 71-87 (1997)

[3] Alicki, R. and Fannes, M., Defining quantum dynamical entropy, 32, 75-82 (1994).

[4] Araki, H., Relative entropy for states of von Neumann algebras, Publ. RIMS Kyoto Univ., 11, 809-833 (1976).

[5] Araki, H., Relative entropy for states of von Neumann algebras II, 13, 173-192 (1977).

[6] Barnum, H., Nielsen, M.A., and Schumacher, B.W., Information transmission through a noisy quantum channel, Physical Review A, 57, No.6, 4153-4175 (1998).

[7] Belavkin, V.P., and Ohya, M., Quantum entropy and information in discrete entangled states, Infinite Dimensional Analysys, Quantum Probability and Related Topics, 4, No.2, 137-160 (2001).

[8] Belavkin, V.P., and Ohya, M., Entanglement, quantum entropy and mutual information, Proceedings of the Royal Society of London, Series A, Mathematical and Physical Sciences, 458, 209-231 (2002).

[9] Benatti, F., Trieste Notes in Phys, Springer-Verlag, (1993).

[10] Bennett, C.H., Shor, P.W., Smolin, J.A., and Thapliyalz, A.V., Entanglement-Assisted Capacity of a Quantum Channel and the Reverse Shannon Theorem, quant-ph/0106052.

[11] Choda, M., Entropy for extensions of Bernoulli shifts, 16, 1197-1206 (1996).

[12] Connes, A., Narnhofer, H. and Thirring, W., Dynamical entropy of C*-algebras and von Neumann algebras, 112, 691-719 (1987).

[13] Connes, A., and Størmer, E., Entropy for automorphisms of von Neumann algebras, Acta Math., 134, 289-306 (1975).

[14] Donald, M. J., On the relative entropy, 105, 13-34 (1985).

[15] Emch, G.G., Positivity of the K-entropy on non-abelian K-flows, Z. Wahrscheinlichkeitstheoryverw. Gebiete, 29, 241 (1974).

[16] Fichtner, K.H., Freudenberg, W., and Liebscher, V., Beam splittings and time evolutions of Boson systems, Fakultat fur Mathematik und Informatik, Math/ Inf/96/ 39, Jena, 105 (1996).

[17] Gelfand, I. M. and Yaglom, A. M., Calculation of the amount of information about a random function contained in another such function, 12, 199-246 (1959).

[18] Holevo, A.S., Some estimates for the amount of information transmittable by a quantum communication channel (in Russian), ProblemyPeredachiInformacii, 9, 3-11 (1973).

[19] Hudetz, T., Topological entropy for appropriately approximated C*-algebras, 35, 4303-4333 (1994).

[20] Ingarden, R. S., Quantum information theory, 10, 43-73 (1976).

[21] Ingarden, R.S., Kossakowski, A., and Ohya, M., Information Dynamics and Open Systems, Kluwer, (1997).

[22] Jamiołkowski, A., Linear transformations which preserve trace and positive semidefiniteness of operators, Rep. Math. Phys. 3, 275 -278 (1972).

[23] Kolmogorov, A. N., Theory of transmission of information, Ser.2, 33, 291-321 (1963).

[24] Kossakowski, A. and Ohya, M., New scheme of quantum teleportation, Infinite Dimensional Analysis, Quantum Probability and Related Topics, 10, No.3, 411-420 (2007).

[25] Kossakowski, A., Ohya, M. and Watanabe, N., Quantum dynamical entropy for completely positive map, Infinite Dimensional Analysis, Quantum Probability and Related Topics, 2, No.2, 267-282, (1999)

[26] Kullback, S. and Leibler, R., On information and sufficiency, 22, 79-86 (1951).

[27] Muraki, N., and Ohya, M., Entropy functionals of Kolmogorov Sinai type and their limit theorems, Letter in Math. Phys., 36, 327-335 (1996)

[28] Muraki, N., Ohya, M., and Petz, D., Entropies of general quantum states, Open Systems and Information Dynamics, 1, 43-56 (1992)

[29] von Neumann, J., Die MathematischenGrundlagen der Quantenmechanik, Springer-Berlin (1932)

[30] Ohya, M., Quantum ergodic channels in operator algebras, J. Math. Anal. Appl., 84, 318-328 (1981)

[31] Ohya, M., On compound state and mutual information in quantum information theory, IEEE Trans. Information Theory, 29, 770-774 (1983)

[32] Ohya, M., Note on quantum probability, L. NuovoCimento, 38, 402-404 (1983).

[33] Ohya, M., Entropy transmission in C*-dynamical systems, J. Math. Anal. Appl., 100, No.1, 222-235 (1984).

[34] Ohya, M., State change and entropies in quantum dynamical systems, 1136, 397-408 (1985).

[35] Ohya, M., Some aspects of quantum information theory and their applications to irreversible processes, Rep. Math. Phys., 27, 19-47 (1989)

[36] Ohya, M., Information dynamics and its application to optical communication processes, Springer Lecture Note in Physics, 378, 81-92, (1991)

[37] Ohya, M., State change, complexity and fractal in quantum systems, Quantum Communications and Measurement, 2, 309-320 (1995)

[38] Ohya, M. Fundamentals of quantum mutual entropy and capacity, 6, 69-78 (1999).

[39] Ohya, M., and Petz, D., Quantum Entropy and its Use, Springer, Berlin, 1993.

[40] Ohya, M., Petz, D., and Watanabe, N., On capacity of quantum channels, Probability and Mathematical Statistics, 17, 179-196 (1997)

[41] Ohya, M., Petz, D., and Watanabe, N., Numerical computation of quantum capacity, International Journal of Theoretical Physics, 37, No.1, 507-510 (1998).

[42] Ohya, M. and Volovich, I. V. On Quantum Entropy and its Bound, 6, 301 - 310 (2003).

[43] Ohya, M., and I. Volovich, I., Mathematical Foundations of Quantum Information. and Computation and Its Applications to Nano- and Bio-systems, Springer, (2011).

[44] Ohya, M., and Watanabe, N., Construction and analysis of a mathematical model in quantum communication processes, Electronics and Communications in Japan, Part 1, 68, No.2, 29-34 (1985)

[45] Ohya, M., and Watanabe, N., Quantum capacity of noisy quantum channel, Quantum Communication and Measurement, 3, 213-220 (1997)

[46] Ohya, M., and Watanabe, N., Foundation of Quantum Communication Theory (in Japanese), Makino Pub. Co., (1998)

[47] Ohya, M., and Watanabe, N., Comparison of mutual entropy - type measures, TUS preprint (2003).

[48] Park, Y. M., Dynamical entropy of generalized quantum Markov chains, 32, 63-74 (1994).

[49] Schatten, R., Norm Ideals of Completely Continuous Operators, Springer-Verlag, (1970).

[50] Schumacher, B.W., Sending entanglement through noisy quantum channels, Physical Review A, 54, 2614 (1996).

[51] Schumacher, B.W., and Nielsen, M.A., Quantum data processing and error correction, Physical Review A, 54, 2629 (1996).

[52] Shannon, C. E., A mathematical theory of communication, 27, 379 - 423 and 623 - 656.

[53] Shor, P., The quantum channel capacity and coherent information, Lecture Notes, MSRI Workshop on Quantum Computation, (2002).

[54] Uhlmann, A., Relative entropy and the Wigner-Yanase-Dyson-Lieb concavity in interpolation theory, Commun. Math. Phys., 54, 21-32 (1977)

[55] Umegaki, H., Conditional expectations in an operator algebra IV (entropy and information), Kodai Math. Sem. Rep., 14, 59-85 (1962).

[56] Urbanik, K., Joint probability distribution of observables in quantum mechanics, 21, 117-133 (1961).

[57] Voiculescu, D., Dynamical approximation entropies and topological entropy in operator algebras, 170, 249-281 (1995).

[58] Watanabe, N., Some aspects of complexities for quantum processes, Open Systems and Information Dynamics, 16, 293-304 (2009).

[59] Watanabe, N., Note on entropies of quantum dynamical systems, Foundation of Physics, 41, 549-563 DOI 10, 1007/s10701-010-9455-x (2011).

Section 2

Quantum Optics - Some Complex Aspects

Chapter 5

Information Theory and Entropies for Quantized Optical Waves in Complex Time-Varying Media

Jeong Ryeol Choi

Additional information is available at the end of the chapter

1. Introduction

An important physical intuition that led to the Copenhagen interpretation of quantum mechanics is the Heisenberg uncertainty relation (HUR) which is a consequence of the noncommutativity between two conjugate observables. Our ability of observation is intrinsically limited by the HUR, quantifying an amount of inevitable and uncontrollable disturbance on measurements (Ozawa, 2004).

Though the HUR is one of the most fundamental results of the whole quantum mechanics, some drawbacks concerning its quantitative formulation are reported. As the expectation value of the commutator between two arbitrary noncommuting operators, the value of the HUR is not a fixed lower bound and varies depending on quantum state (Deutsch, 1983). Moreover, in some cases, the ordinary measure of uncertainty, i.e., the variance of canonical variables, based on the Heisenberg-type formulation is divergent (Abe et al., 2002).

These shortcommings are highly nontrivial issues in the context of information sciences. Thereby, the theory of informational entropy is proposed as an alternate optimal measure of uncertainty. The adequacy of the entropic uncertainty relations (EURs) as an uncertainty measure is owing to the fact that they only regard the probabilities of the different outcomes of a measurement, whereas the HUR the variances of the measured values themselves (Werner, 2004). According to Khinchin's axioms (Ash, 1990) for the requirements of common information measures, information measures should be dependent exclusively on a probability distribution (Pennini & Plastino, 2007). Thank to active research and technological progress associated with quantum information theory (Nielsen & Chuang, 2000; Choi et al., 2011), the entropic uncertainty band now became a new concept in quantum physics.

Information theory proposed by Shannon (Shannon, 1948a; Shannon, 1948b) is important as information-theoretic uncertainty measures in quantum physics but even in other areas such as signal and/or image processing. Essential unity of overall statistical information for a system can be demonstrated from the Shannon information, enabling us to know how information could be quantified with absolute precision. Another good measure of uncertainty or randomness is the Fisher information (Fisher, 1925) which appears as the basic ingredient in bounding entropy production. The Fisher information is a measure of accuracy in statistical theory and useful to estimate ultimate limits of quantum measurements.

Recently, quantum information theory besides the fundamental quantum optics has aroused great interest due to its potential applicability in three sub-regions which are quantum computation, quantum communication, and quantum cryptography. Information theory has contributed to the development of the modern quantum computation (Nielsen & Chuang, 2000) and became a cornerstone in quantum mechanics. A remarkable ability of quantum computers is that they can carry out certain computational tasks exponentially faster than classical computers utilizing the entanglement and superposition principle.

Stimulated by these recent trends, this chapter is devoted to the study of information theory for optical waves in complex time-varying media with emphasis on the quantal information measures and informational entropies. Information theoretic uncertainty relations and the information measures of Shannon and Fisher will be managed. The EUR of the system will also be treated, quantifying its physically allowed minimum value using the invariant operator theory established by Lewis and Riesenfeld (Lewis, 1967; Lewis & Riesenfeld, 1969). Invariant operator theory is crucial for studying quantum properties of complicated time-varying systems, since it, in general, gives exact quantum solutions for a system described by time-dependent Hamiltonian so far as its counterpart classical solutions are known.

2. Quantum optical waves in time-varying media

Let us consider optical waves propagating through a linear medium that has time-dependent electromagnetic parameters. Electromagnetic properties of the medium are in principle determined by three electromagnetic parameters such as electric permittivity ϵ, magnetic permeability μ, and conductivity σ. If one or more parameters among them vary with time, the medium is designated as a time-varying one. Coulomb gauge will be taken for convenience under the assumption that the medium have no net charge distributions. Then the scalar potential vanishes and, consequently, the vector potential is the only potential needed to consider when we develop quantum theory of electromagnetic wave phenomena. Regarding this fact, the quantum properties of optical waves in time-varying media are described in detail in Refs. (Choi & Yeon, 2005; Choi, 2012; Choi et al, 2012) and they will be briefly surveyed in this section as a preliminary step for the study of information theory.

According to separation of variables method, it is favorable to put vector potential in the form

$$\mathbf{A}(\mathbf{r},t) = \sum_l \mathbf{u}_l(\mathbf{r}) q_l(t). \tag{1}$$

Then, considering the fact that the fields and current density obey the relations, $\mathbf{D} = \epsilon(t)\mathbf{E}$, $\mathbf{B} = \mu(t)\mathbf{H}$, and $\mathbf{J} = \sigma(t)\mathbf{E}$, in linear media, we derive equation of motion for q_l from Maxwell equations as (Choi, 2012; Choi, 2010a; Pedrosa & Rosas, 2009)

$$\ddot{q}_l + \{[\dot{\epsilon}(t) + \sigma(t)]/\epsilon(t)\}\dot{q}_l + \omega_l^2(t) q_l = 0. \tag{2}$$

Here, the angular frequency (natural frequency) is given by $\omega_l(t) = c(t)k_l$ where $c(t)$ is the speed of light in media and $k_l (= |\mathbf{k}_l|)$ is the wave number. Because electromagnetic parameters vary with time, $c(t)$ can be represented as a time-dependent form, i.e., $c(t) = 1/\sqrt{\mu(t)\epsilon(t)}$. However, $k_l (= |\mathbf{k}_l|)$ is constant since it does not affected by time-variance of the parameters.

The formula of mode function $\mathbf{u}_l(\mathbf{r})$ depends on the geometrical boundary condition in media (Choi & Yeon, 2005). For example, it is given by $\mathbf{u}_{l\nu}(\mathbf{r}) = V^{-1/2}\hat{\varepsilon}_{l\nu} \exp(\pm i\mathbf{k}_l \cdot \mathbf{r})$ $(\nu = 1, 2)$ for the fields propagating under the periodic boundary condition, where V is the volume of the space, $\hat{\varepsilon}_{l\nu}$ is a unit vector in the direction of polarization designated by ν.

From Hamilton's equations of motion, $\dot{q}_l = \partial H_l/\partial p_l$ and $\dot{p}_l = -\partial H_l/\partial q_l$, the classical Hamiltonian that gives Eq. (3) can be easily established. Then, by converting canonical variables, q_l and p_l, into quantum operators, $\hat{q}_l$ and $\hat{p}_l$, from the resultant classical Hamiltonian, we have the quantum Hamiltonian such that (Choi et al., 2012)

$$\hat{H}_l(\hat{q}_l, \hat{p}_l, t) = \frac{1}{2\epsilon_0} e^{-\Lambda(t)} \hat{p}_l^2 + b(t)(\hat{q}_l\hat{p}_l + \hat{p}_l\hat{q}_l) + \frac{1}{2}\epsilon_0 e^{\Lambda(t)} \varpi_l^2(t)\hat{q}_l^2, \tag{3}$$

where $\hat{p}_l = -i\hbar(\partial/\partial q_l)$, $\epsilon_0 = \epsilon(0)$, $b(t)$ is an arbitrary time function, and

$$\Lambda(t) = \int_0^t dt' [\dot{\epsilon}(t') + \sigma(t')]/\epsilon(t'), \tag{4}$$

$$\varpi_l^2(t) = \omega_l^2(t) + 2\dot{b}(t) + 2b(t)[\dot{\epsilon}(t) + \sigma(t)]/\epsilon(t) + 4b^2(t). \tag{5}$$

The complete Hamiltonian is obtained by summing all individual Hamiltonians: $\hat{H} = \sum_l \hat{H}_l(\hat{q}_l, \hat{p}_l, t)$.

From now on, let us treat the wave of a particular mode and drop the under subscript l for convenience. It is well known that quantum problems of optical waves in nonstationary media are described in terms of classical solutions of the system. Some researchers use real classical solutions (Choi, 2012; Pedrosa & Rosas, 2009) and others imaginary solutions (Angelow & Trifonov, 2010; Malkin et al., 1970). In this chapter, real solutions of classical equation of motion for q will be considered. Since Eq. (2) is a second order differential equation, there are two linearly independent classical solutions. Let us denote them as $s_1(t)$ and $s_2(t)$, respectively. Then, we can define an Wronskian of the form

$$\Omega = 2\epsilon_0 e^{\Lambda(t)} \left[s_1(t)\frac{ds_2(t)}{dt} - \frac{ds_1(t)}{dt}s_2(t) \right]. \tag{6}$$

This will be used at later time, considering only the case that $\Omega > 0$ for convenience.

When we study quantum problem of a system that is described by a time-dependent Hamiltonian such as Eq. (3), it is very convenient to introduce an invariant operator of the system. Such idea (invariant operator method) is firstly devised by Lewis and Riesenfeld (Lewis, 1967; Lewis & Riesenfeld, 1969) in a time-dependent harmonic oscillator as mentioned in the introductory part and now became one of potential tools for investigating quantum characteristics of time-dependent Hamiltonian systems. By solving the Liouville-von Neumann equation of the form

$$\frac{d\hat{K}}{dt} = \frac{\partial \hat{K}}{\partial t} + \frac{1}{i\hbar}[\hat{K}, \hat{H}] = 0, \tag{7}$$

we obtain the invariant operator of the system as (Choi, 2004)

$$\hat{K} = \hbar\Omega\left(\hat{a}^\dagger\hat{a} + \frac{1}{2}\right), \tag{8}$$

where Ω is chosen to be positive from Eq. (6) and $\hat{a}$ and $\hat{a}^\dagger$ are annihilation and creation operators, respectively, that are given by

$$\hat{a} = \sqrt{\frac{1}{\hbar\Omega}}\left\{\left[\frac{\Omega}{2s(t)} - i\epsilon_0 e^{\Lambda(t)}\left(\frac{ds(t)}{dt} - 2b(t)s(t)\right)\right]\hat{q} + is(t)\hat{p}\right\}, \tag{9}$$

$$\hat{a}^\dagger = \sqrt{\frac{1}{\hbar\Omega}}\left\{\left[\frac{\Omega}{2s(t)} + i\epsilon_0 e^{\Lambda(t)}\left(\frac{ds(t)}{dt} - 2b(t)s(t)\right)\right]\hat{q} - is(t)\hat{p}\right\}, \tag{10}$$

with

$$s(t) = \sqrt{s_1^2(t) + s_2^2(t)}. \tag{11}$$

Since the system is somewhat complicate, let us develop our theory with $b(t) = 0$ from now on. Then, Eq. (5) just reduces to $\varpi_l(t) = \omega_l(t)$. Since the formula of Eq. (8) is very similar to the familiar Hamiltonian of the simple harmonic oscillator, we can solve its eigenvalue equation via well known conventional method. The zero-point eigenstate $\phi_0(q,t)$ of $\hat{K}$ is obtained from $\hat{a}\phi_0(q,t) = 0$. Once $\phi_0(q,t)$ is obtained, nth eigenstates are also derived by acting $\hat{a}^\dagger$ on $\phi_0(q,t)$ n times. Hence we finally have (Choi, 2012)

$$\phi_n(q,t) = \sqrt[4]{\frac{\delta(t)}{\pi}}\frac{1}{\sqrt{2^n n!}}H_n\left(\sqrt{\delta(t)}q\right) \times \exp\left\{-\frac{\delta(t)}{2}\left[1 - i\frac{2\epsilon_0 e^{\Lambda(t)}s(t)}{\Omega}\frac{ds(t)}{dt}\right]q^2\right\}, \tag{12}$$

where $\delta(t) = \Omega/[2\hbar s^2(t)]$ and H_n are Hermite polynomials.

According to the theory of Lewis-Riesenfeld invariant, the wave functions that satisfy the Schrödinger equation are given in terms of $\phi_n(q,t)$:

$$\psi_n(q,t) = \phi_n(q,t)\exp[i\theta_n(t)], \tag{13}$$

where $\theta_n(t)$ are time-dependent phases of the wave functions. By substituting Eq. (13) with Eq. (3) into the Schrödinger equation, we derive the phases to be $\theta_n(t) = -(n+1/2)\,\eta(t)$ where (Choi, 2012)

$$\eta(t) = \frac{\Omega}{2\epsilon_0}\int_0^t \frac{dt'}{s^2(t')e^{\Lambda(t')}} + \eta(0). \tag{14}$$

The probability densities in both q and p spaces are given by the square of wave functions, i.e., $\rho_n(q) = |\psi_n(q,t)|^2$ and $\tilde{\rho}_n(p) = |\tilde{\psi}_n(p,t)|^2$, respectively. From Eq. (13) and its Fourier component, we see that

$$\rho_n(q) = \sqrt{\frac{\delta(t)}{\pi}}\frac{1}{2^n n!}\left\{H_n[\sqrt{\delta(t)}q]\right\}^2 e^{-\delta(t)q^2}, \tag{15}$$

$$\tilde{\rho}_n(p) = \sqrt{\frac{\delta'(t)}{\pi}}\frac{1}{2^n n!}\left\{H_n[\sqrt{\delta'(t)}p]\right\}^2 e^{-\delta'(t)p^2}, \tag{16}$$

where

$$\delta'(t) = \frac{2\Omega}{\hbar\left[\frac{\Omega^2}{s^2(t)} + 4\epsilon_0^2 e^{2\Lambda(t)}\left(\frac{ds(t)}{dt}\right)^2\right]}. \tag{17}$$

The wave functions and the probability densities derived here will be used in subsequent sections in order to develop the information theory of the system.

3. Information measures for thermalized quantum optical fields

Informations of a physical system can be obtained from the statistical analysis of results of a measurement performed on it. There are two important information measures. One is the Shannon information and the other is the Fisher information. The Shannon information is also called as the Wehrl entropy in some literatures (Wehrl, 1979; Pennini & Plastino, 2004) and suitable for measuring uncertainties relevant to both quantum and thermal effects whereas quantum effect is overlooked in the concept of ordinary entropy. The Fisher information which is also well known in the field of information theory provides the extreme physical information through a potential variational principle. To manage these informations, we start from the establishment of density operator for the electromagnetic field equilibrated with its environment of temperature T. Density operator of the system obeys the Liouville-von Neumann equation such that (Choi et al., 2011)

$$\frac{\partial\hat{\varrho}(t)}{\partial t} + \frac{1}{i\hbar}[\hat{\varrho}(t),\ \hat{H}] = 0. \tag{18}$$

Considering the fact that invariant operator given in Eq. (8) is also established via the Liouville-von Neumann equation, we can easily construct density operator as a function of the invariant operator. This implies that the Hamiltonian $\hat{H}$ in the density operator of the simple harmonic oscillator should be replaced by a function of the invariant operator $y(0)\hat{\mathcal{K}}$, where $y(0)\{= [2\epsilon_0 e^{\Lambda(0)} s^2(0)]^{-1}\}$ is inserted for the purpose of dimensional consideration. Thus we have the density operator in the form

$$\hat{\varrho}(t) = \frac{1}{Z} e^{-\beta\hbar W(\hat{a}^\dagger \hat{a} + 1/2)}, \tag{19}$$

where $W = y(0)\Omega$, Z is a partition function, $\beta = k_b T$, and k_b is Boltzmann's constant. If we consider Fock state expression, the above equation can be expand to be

$$\varrho(t) = \frac{1}{Z} \sum_{n=0}^{\infty} |\phi_n(t)\rangle e^{-\beta\hbar W(n+1/2)} \langle \phi_n(t)|, \tag{20}$$

while the partition function becomes

$$Z = \sum_{n=0}^{\infty} \langle \phi_n(t)| e^{-\beta\hbar W(\hat{a}^\dagger \hat{a} + 1/2)} |\phi_n(t)\rangle. \tag{21}$$

If we consider that the coherent state is the most classical-like quantum state, a semiclassical distribution function associated with the coherent state may be useful for the description of information measures. As is well known, the coherent state $|\alpha\rangle$ is obtained by solving the eigenvalue equation of $\hat{a}$:

$$\hat{a}|\alpha\rangle = \alpha|\alpha\rangle. \tag{22}$$

Now we introduce the semiclassical distribution function $\mu_\varrho(\alpha)$ related with the density operator via the equation (Anderson & Halliwell, 1993)

$$\mu_\varrho(\alpha) = \langle \alpha | \varrho(t) | \alpha \rangle. \tag{23}$$

This is sometimes referred to as the Husimi distribution function (Husimi, 1940) and appears frequently in the study relevant to the Wigner distribution function for thermalized quantum systems. The Wigner distribution function is regarded as a qusi-distribution function because some parts of it are not positive but negative. In spite of the ambiguity in interpreting this negative value as a distribution function, the Wigner distribution function meets all requirements of both quantum and statistical mechanics, i.e., it gives correct expectation values of quantum mechanical observables. In fact, the Husimi distribution function can also be constructed from the Wigner distribution function through a mathematical procedure known as "Gaussian smearing" (Anderson & Halliwell, 1993). Since this smearing washes out the negative part, the negativity problem is resolved by the Hisimi's work. But it is interesting to note that new drawbacks are raised in that case, which state that the probabilities of different samplings of q and p, relevant to the Husimi distribution function, cannot be represented by mutually exclusive states due to the lack of orthogonality of coherent states used for example in Eq. (23) (Anderson & Halliwell, 1993; Nasiri, 2005). This weak point is however almost fully negligible in the actual study of the system, allowing us to use the Husimi distribution function as a powerful means in the realm of semiclassical statistical physics.

Notice that coherent state can be rewritten in the form

$$|\alpha\rangle = \hat{D}(\alpha)|\phi_0(t)\rangle, \tag{24}$$

where $\hat{D}(\alpha)$ is the displacement operator of the form $\hat{D}(\alpha) = e^{\alpha\hat{a}^\dagger - \alpha^*\hat{a}}$. A little algebra leads to

$$|\alpha\rangle = \exp\left(-\frac{1}{2}|\alpha|^2\right)\sum_n \frac{\alpha^n}{\sqrt{n!}}|\phi_n(t)\rangle. \tag{25}$$

Hence, the coherent state is expanded in terms of Fock state wave functions. Now using Eqs. (20) and (25), we can evaluate Eq. (23) to be

$$\begin{aligned} \mu_\varrho(\alpha) &= \frac{1}{Z}\sum_{n=0}^{\infty} e^{-\beta\hbar W(n+1/2)}|\langle\phi_n(t)|\alpha\rangle|^2 \\ &= \frac{1-e^{-\beta\hbar W}}{\exp[(1-e^{-\beta\hbar W})|\alpha|^2]}. \end{aligned} \tag{26}$$

Here, we used a well known relation in photon statistics, which is

$$|\langle\phi_n(t)|\alpha\rangle|^2 = \frac{|\alpha|^{2n}}{n!}e^{-|\alpha|^2}. \tag{27}$$

As you can see, the Husimi distribution function is strongly related to the coherent state and it provides necessary concepts for establishment of both the Shannon and the Fisher informations. If we consider Eqs. (9) and (22), α (with $b(t) = 0$) can be written as

$$\alpha = \sqrt{\frac{1}{\hbar\Omega}}\left\{\left[\frac{\Omega}{2s(t)} + i\epsilon_0 e^{\Lambda(t)}\frac{ds(t)}{dt}\right]q + is(t)p\right\}. \tag{28}$$

Hence there are innumerable number of α-samples that correspond to different pair of (q,p), which need to be examined for measurement.

A natural measure of uncertainty in information theory is the Shannon information as mentioned earlier. The Shannon information is defined as (Anderson & Halliwell, 1993)

$$I_S = -\int \frac{d^2\alpha}{\pi}\mu_\varrho(\alpha)\ln\mu_\varrho(\alpha), \tag{29}$$

where $d^2\alpha = d\mathrm{Re}(\alpha)\, d\mathrm{Im}(\alpha)$. With the use of Eq. (26), we easily derive it:

$$I_S = 1 + \ln \frac{1}{1 - e^{-\beta\hbar W}}. \tag{30}$$

This is independent of time and, in the limiting case of fields propagating in time-*in*dependent media that have no conductivity, W becomes natural frequency of light, leading this formula to correspond to that of the simple harmonic oscillator (Pennini & Plastino, 2004). This approaches to $I_S \simeq \ln[k_b T/(\hbar W)]$ for sufficiently high temperature, yielding the dominance of the thermal fluctuation and, consequently, permitting the quantum fluctuation to be neglected. On the other hand, as T decreases toward absolute zero, the Shannon information is always larger than unity ($I_S \geq 1$). This condition is known as the Lieb-Wehrl condition because it is conjectured by Wehrl (Wehrl, 1979) and proved by Lieb (Lieb, 1978). From this we can see that I_S has a lower bound which is connected with pure quantum effects. Therefore, while usual entropy is suitable for a measure of uncertainty originated only from thermal fluctuation, I_S plays more universal uncertainty measure covering both thermal and quantum regimes (Anderson & Halliwell, 1993).

Other potential measures of information are the Fisher informations which enable us to assess intrinsic accuracy in the statistical estimation theory. Let us consider a system described by the stochastic variable $\alpha = \alpha(x)$ with a physical parameter x. When we describe a measurement of α in order to infer x from the measurement, it is useful to introduce the coherent-state-related Fisher information that is expressed in the form

$$I_{F,x} = \int \frac{d^2\alpha}{\pi} f(\alpha(x);x) \left(\frac{\partial \ln f(\alpha(x);x)}{\partial x} \right)^2. \tag{31}$$

In fact, there are many different scenarios of this information depending on the choice of x. For a more general definition of the Fisher information, you can refer to Ref. (Pennini & Plastino, 2004).

If we take $f(\alpha(x);x) = \mu_\varrho(\alpha)$ and $x = \beta$, the Fisher's information measure can be written as (Pennini & Plastino, 2004)

$$I_{F,\beta} = \int \frac{d^2\alpha}{\pi} \mu_\varrho(\alpha) \left(\frac{\partial \ln \mu_\varrho(\alpha)}{\partial \beta} \right)^2. \tag{32}$$

Since β is the parameter to be estimated here, I_β reflects the change of μ_ϱ according to the variation of temperature. A straightforward calculation yields

$$I_{F,\beta} = \left(\frac{\hbar W}{e^{\beta\hbar W} - 1} \right)^2. \tag{33}$$

This is independent of time and of course agree, in the limit of the simple harmonic oscillator case, to the well known formula of Pennini and Plastino (Pennini & Plastino, 2004). Hence, the change of electromagnetic parameters with time does not affect to the value of β. $I_{F,\beta}$ reduces to zero at absolute zero-temperature ($T \to 0$), leading to agreement with the third law of thermodynamics (Pennini & Plastino, 2007).

Another typical form of the Fisher informations worth to be concerned is the one obtained with the choice of $f(\alpha(x);x) = \mu_\varrho(\alpha)$ and $x = \{q,p\}$ (Pennini et al, 1998):

$$I_{F,\{q,p\}} = \int \frac{d^2\alpha}{\pi}\mu_\varrho(\alpha)\left[\sigma_{qq,\alpha}\left(\frac{\partial \ln \mu_\varrho(\alpha)}{\partial q}\right)^2 + \sigma_{pp,\alpha}\left(\frac{\partial \ln \mu_\varrho(\alpha)}{\partial p}\right)^2\right], \tag{34}$$

where $\sigma_{qq,\alpha}$ and $\sigma_{pp,\alpha}$ are variances of q and p in the Glauber coherent state, respectively. Notice that $\sigma_{qq,\alpha}$ and $\sigma_{pp,\alpha}$ are inserted here in order to consider the weight of two independent terms in Eq. (34). As you can see, this information is jointly determined by means of canonical variables q and p. To evaluate this, we need

$$\sigma_{qq,\alpha} = \langle\alpha|\hat{q}^2|\alpha\rangle - \langle\alpha|\hat{q}|\alpha\rangle^2, \tag{35}$$

$$\sigma_{pp,\alpha} = \langle\alpha|\hat{p}^2|\alpha\rangle - \langle\alpha|\hat{p}|\alpha\rangle^2. \tag{36}$$

It may be favorable to represent $\hat{q}$ and $\hat{p}$ in terms of $\hat{a}$ and $\hat{a}^\dagger$ at this stage. They are easily obtained form the inverse representation of Eqs. (9) and (10) to be

$$\hat{q} = \sqrt{\hbar/\Omega}s(t)[\hat{a} + \hat{a}^\dagger], \tag{37}$$

$$\hat{p} = \sqrt{\hbar}\left[\left(\frac{\epsilon_0 e^{\Lambda(t)}}{\sqrt{\Omega}}\frac{ds(t)}{dt} - i\frac{\sqrt{\Omega}}{2s(t)}\right)\hat{a} + \left(\frac{\epsilon_0 e^{\Lambda(t)}}{\sqrt{\Omega}}\frac{ds(t)}{dt} + i\frac{\sqrt{\Omega}}{2s(t)}\right)\hat{a}^\dagger\right]. \tag{38}$$

Thus with the use of these, Eqs. (35) and (36) become

$$\sigma_{qq,\alpha} = \frac{\hbar s^2(t)}{\Omega}, \tag{39}$$

$$\sigma_{pp,\alpha} = \frac{\hbar\Omega}{4s^2(t)}\left[1 + 4\epsilon_0^2 e^{2\Lambda(t)}\frac{s^2(t)}{\Omega^2}\left(\frac{ds(t)}{dt}\right)^2\right]. \tag{40}$$

A little evaluation after substituting these quantities into Eq. (34) leads to

$$I_{F,\{q,p\}} = \left[1 + 4\epsilon_0^2 e^{2\Lambda(t)}\frac{s^2(t)}{\Omega^2}\left(\frac{ds(t)}{dt}\right)^2\right](1 - e^{-\beta\hbar W}). \tag{41}$$

Notice that this varies depending on time. In case that the time dependence of every electromagnetic parameters vanishes and $\sigma \to 0$, this reduces to that of the simple harmonic oscillator limit, $I_{F,\{q,p\}} = 1 - e^{-\beta\hbar\omega}$, where natural frequency ω is constant, which exactly agrees with the result of Pennini and Plastino (Pennini & Plastino, 2004).

4. Husimi uncertainties and uncertainty relations

Uncertainty principle is one of intrinsic features of quantum mechanics, which distinguishes it from classical mechanics. Aside form conventional procedure to obtain uncertainty relation, it may be instructive to compute a somewhat different uncertainty relation for optical waves through a complete mathematical description of the Husimi distribution function. Bearing in mind this, let us see the uncertainty of canonical variables, associated with information measures, and their corresponding uncertainty relation. The definition of uncertainties suitable for this purpose are

$$\sigma_{\mu,qq}(t) = \langle \hat{q}^2 \rangle_\mu - \langle \hat{q} \rangle_\mu^2, \tag{42}$$

$$\sigma_{\mu,pp}(t) = \langle \hat{p}^2 \rangle_\mu - \langle \hat{p} \rangle_\mu^2, \tag{43}$$

$$\sigma_{\mu,qp}(t) = \langle \hat{q}\hat{p} + \hat{p}\hat{q} \rangle_\mu /2 - \langle \hat{q} \rangle_\mu \langle \hat{p} \rangle_\mu, \tag{44}$$

where $\langle \hat{O}^l \rangle_\mu$ $(l = 1, 2)$ with an arbitrary operator $\hat{O}$ is the expectation value relevant to the Husimi distribution function and can be evaluated from

$$\langle \hat{O}^l \rangle_\mu = \int \frac{d^2\alpha}{\pi} O^l \mu_\varrho(\alpha). \tag{45}$$

While $\langle \hat{q} \rangle_\mu = 0$ and $\langle \hat{p} \rangle_\mu = 0$ for $l = 1$, the rigorous algebra for higher orders give

$$\langle \hat{q}^2 \rangle_\mu = \frac{2\hbar s^2(t)}{\Omega} R(\beta), \tag{46}$$

$$\langle \hat{p}^2 \rangle_\mu = \frac{\hbar\Omega}{2s^2(t)} \left[1 + 4\epsilon_0^2 e^{2\Lambda(t)} \frac{s^2(t)}{\Omega^2} \left(\frac{ds(t)}{dt} \right)^2 \right] R(\beta), \tag{47}$$

$$\langle \hat{q}\hat{p} + \hat{p}\hat{q} \rangle_\mu = 4\hbar\epsilon_0 e^{\Lambda(t)} \frac{s(t)}{\Omega} \frac{ds(t)}{dt} R(\beta), \tag{48}$$

where

$$R(\beta) = \frac{1}{1 - e^{-\beta\hbar W}} + \frac{1}{2}. \tag{49}$$

Thus we readily have

$$\sigma_{\mu,qq} = \frac{2\hbar s^2(t)}{\Omega} R(\beta), \tag{50}$$

$$\sigma_{\mu,pp} = \frac{\hbar\Omega}{2s^2(t)} \left[1 + 4\epsilon_0^2 e^{2\Lambda(t)} \frac{s^2(t)}{\Omega^2} \left(\frac{ds(t)}{dt} \right)^2 \right] R(\beta). \tag{51}$$

Like other types of uncertainties in physics, the relationship between $\sigma_{\mu,qq}$ and $\sigma_{\mu,pp}$ is rather unique, i.e., if one of them become large the other become small, and there is nothing whatever one can do about it.

We can represent the uncertainty product σ_μ and the generalized uncertainty product Σ_μ in the form

$$\sigma_\mu = [\sigma_{\mu,qq}(t)\sigma_{\mu,pp}(t)]^{1/2}, \tag{52}$$

$$\Sigma_\mu = [\sigma_{\mu,qq}(t)\sigma_{\mu,pp}(t) - \sigma_{\mu,qp}{}^2(t)]^{1/2}. \tag{53}$$

Through the use of Eqs. (50) and (51), we get

$$\sigma_\mu = \hbar \left[1 + 4\epsilon_0^2 e^{2\Lambda(t)} \frac{s^2(t)}{\Omega^2} \left(\frac{ds(t)}{dt}\right)^2\right]^{1/2} R(\beta), \tag{54}$$

$$\Sigma_\mu = \hbar R(\beta). \tag{55}$$

Notice that σ_μ varies depending on time, while Σ_μ does not and is more simple form. The relationship between σ_μ and usual thermal uncertainty relations σ obtained using the method of thermofield dynamics (Choi, 2010b; Leplae et al., 1974) are given by $\sigma_\mu = r(\beta)\sigma$ where $r(\beta) = (3e^{\beta\hbar W} - 1)/(e^{\beta\hbar W} + 1)$.

5. Entropies and entropic uncertainty relations

The HUR is employed in many statistical and physical analyses of optical data measured from experiments. This is a mathematical outcome of the nonlocal Fourier analysis (Bohr, 1928) and we can simply represent it by multiplying standard deviations of q and p together. From measurements, simultaneous prediction of q and p with high precision for both beyond certain limits levied by quantum mechanics is impossible according to the Heisenberg uncertainty principle. It is plausible to use the HUR as a measure of the spread when the curve of distribution involves only a simple hump such as the case of Gaussian type. However, the HUR is known to be inadequate when the distribution of the statistical data is somewhat complicated or reveals two or more humps (Bialynicki-Birula, 1984; Majernik & Richterek, 1997).

For this reason, the EUR is suggested as an alternative to the HUR by Bialynicki-Birula and Mycielski (Biatynicki-Birula & Mycielski, 1975). To study the EUR, we start from entropies of q and p associated with the Shannon's information theory:

$$S(\rho_n) = -\int \rho_n(q) \ln \rho_n(q) dq, \tag{56}$$

$$S(\tilde{\rho}_n) = -\int \tilde{\rho}_n(p) \ln \tilde{\rho}_n(p) dp. \tag{57}$$

By executing some algebra after inserting Eqs. (15) and (16) into the above equations, we get

$$S(\rho_n) = -\frac{1}{2} \ln \frac{\Omega}{2\hbar s^2(t)} + \ln(2^n n! \sqrt{\pi}) + n + \frac{1}{2} - \frac{1}{2^n n! \sqrt{\pi}} E(H_n), \tag{58}$$

$$S(\tilde{\rho}_n) = \frac{1}{2}\ln\left\{\frac{\hbar}{2\Omega}\left[\frac{\Omega^2}{s^2(t)} + 4\epsilon_0^2 e^{2\Lambda(t)}\left(\frac{ds(t)}{dt}\right)^2\right]\right\} + \ln(2^n n!\sqrt{\pi}) + n + \frac{1}{2}$$
$$-\frac{1}{2^n n!\sqrt{\pi}}E(H_n), \tag{59}$$

where $E(H_n)$ are entropies of Hermite polynomials of the form (Dehesa et al, 2001)

$$E(H_n) = \int_{-\infty}^{\infty} [H_n(y)]^2 e^{-y^2} \ln([H_n(y)]^2) dy. \tag{60}$$

By adding Eqs. (58) and (59) together,

$$U_{\mathrm{E}} = S(\rho_n) + S(\tilde{\rho}_n), \tag{61}$$

we obtain the alternative uncertainty relation, so-called the EUR such that

$$U_{\mathrm{E}} = \frac{1}{2}\ln\left\{\hbar^2\left[1 + 4\epsilon_0^2 e^{2\Lambda(t)}\frac{s^2(t)}{\Omega^2}\left(\frac{ds(t)}{dt}\right)^2\right]\right\} + 2\ln(2^n n!\sqrt{\pi})$$
$$+2n + 1 - \frac{2}{2^n n!\sqrt{\pi}}E(H_n). \tag{62}$$

The EUR is always larger than or at least equal to a minimum value known as the BBM (Bialynicki-Birula and Mycielski) inequality: $U_{\mathrm{E}} \geq 1 + \ln\pi \simeq 2.14473$ (Haldar & Chakrabarti, 2012). Of course, Eq. (62) also satisfy this inequality. The BBM inequality tells us a lower bound of the uncertainty relation and the equality holds for the case of the simple harmonic oscillation of fields with $n = 0$.

The EUR with evolution in time, as well as information entropy itself, is a potential tool to demonstrate the effects of time dependence of electromagnetic parameters on the evolution of the system and, consequently, it deserves a special interest. The general form of the EUR can also be extended to not only other pairs of observables such as photon number and phase but also more higher dimensional systems even up to infinite dimensions.

6. Application to a special system

The application of the theory developed in the previous sections to a particular system may provide a better understanding of information theory for the system to us. Let us see the case that $\epsilon(t) = \epsilon_0$, $\sigma(t) = 0$, and

$$\mu(t) = \mu_0\frac{1+h}{1+h\cos(\omega_0 t)}, \tag{63}$$

where $\mu_0[=\mu(0)]$, h, and ω_0 are real constants and $|h| \ll 1$. Then, the classical solutions of Eq. (2) are given by

$$s_1(t) = s_0 Ce_\nu(\omega_0 t/2, -\nu h/2), \tag{64}$$

$$s_2(t) = s_0 Se_\nu(\omega_0 t/2, -\nu h/2), \tag{65}$$

where s_0 is a real constant, Ce_ν and Se_ν are Mathieu functions of the cosine and the sine elliptics, respectively, and $\nu = 4k^2/[\epsilon_0\mu_0\omega_0^2(1+h)]$. Figure 1 is information measures for this system, plotted as a function of time. Whereas I_S and $I_{F,\beta}$ do not vary with time, $I_{F,\{q,p\}}$ oscillates as time goes by.

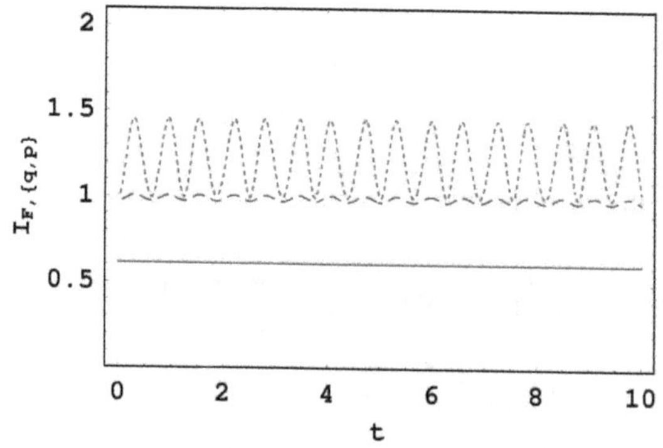

Figure 1. The time evolution of $I_{F,\{q,p\}}$. The values of (k, h) used here are (1, 0.1) for solid red line, (3, 0.1) for long dashed blue line, and (3, 0.2) for short dashed green line. Other parameters are taken to be $\epsilon_0 = 1$, $\mu_0 = 1$, $\beta = 1$, $\hbar = 1$, $\omega_0 = 5$, and $s_0 = 1$.

In case of $h \to 0$, the natural frequency in Eq. (2) become constant and $W \to \omega$. Then, Eqs. (64) and (65) become $s_1 = s_0 \cos \omega t$ and $s_2 = s_0 \sin \omega t$, respectively. We can confirm in this situation that our principal results, Eqs. (30), (33), (41), (54), and (62) reduce to those of the wave described by the simple harmonic oscillator as expected.

7. Summary and conclusion

Information theories of optical waves traveling through arbitrary time-varying media are studied on the basis of invariant operator theory. The time-dependent Hamiltonian that gives classical equation of motion for the time function $q(t)$ of vector potential is constructed. The quadratic invariant operator is obtained from the Liouville-von Neumann equation given in Eq. (7) and it is used as a basic tool for developing information theory of the system. The eigenstates $\phi_n(q,t)$ of the invariant operator are identified using the annihilation and the creation operators. From these eigenstates, we are possible to obtain the Schrödinger solutions, i.e., the wave functions $\psi_n(q,t)$, since $\psi_n(q,t)$ is merely given in terms of $\phi_n(q,t)$.

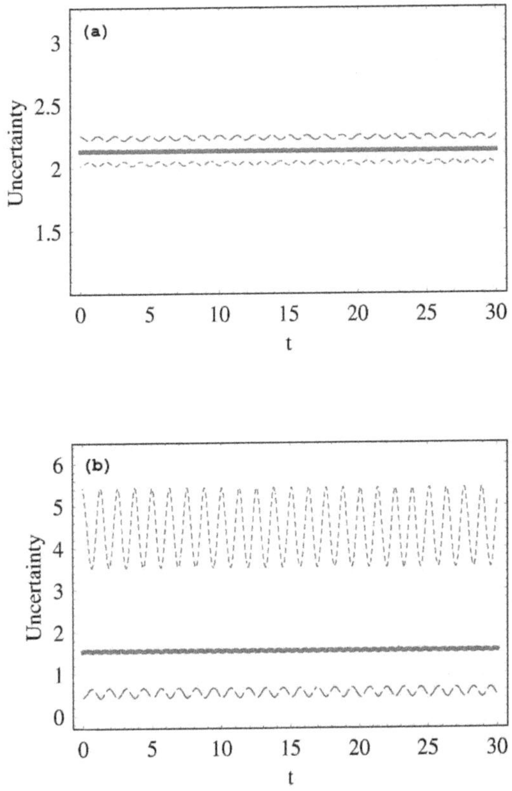

Figure 2. The Uncertainty product σ_μ (thick solid red line) together with $\sigma_{\mu,qq}$ (long dashed blue line) and $\sigma_{\mu,pp}$ (short dashed green line). The values of (k, h) used here are (1, 0.1) for (a) and (3, 0.1) for (b). Other parameters are taken to be $\epsilon_0 = 1$, $\mu_0 = 1$, $\beta = 1$, $\hbar = 1$, $\omega_0 = 5$, and $s_0 = 1$.

The semiclassical distribution function $\mu_Q(\alpha)$ is the expectation value of $\hat{Q}(t)$ in the coherent state which is the very classical-like quantum state. From Eq. (30), we see that the Shannon information does not vary with time. However, Eq. (41) shows that the Fisher information $I_{F,\{q,p\}}$ varies depending on time. It is known that the localization of the density is determined in accordance with the Fisher information (Romera et al, 2005). For this reason, the Fisher measure is regarded as a local measure while the Shannon information is a global information measure of the spreading of density. Local information measures vary depending on various derivatives of the probability density whereas global information measures follow the Kinchin's axiom for information theory (Pennini & Plastino, 2007; Plastino & Casas, 2011).

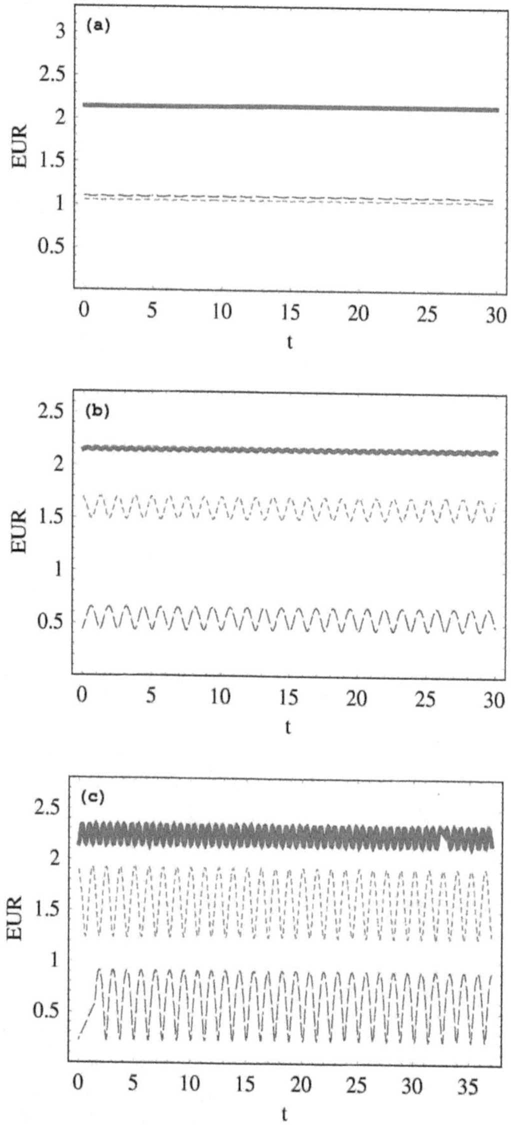

Figure 3. The EUR U_E (thick solid red line) together with $S(\rho_n)$ (long dashed blue line) and $S(\tilde{\rho}_n)$ (short dashed green line). The values of (k, h) used here are (1, 0.1) for (a), (3, 0.1) for (b), and (3, 0.2) for (c). Other parameters are taken to be $\epsilon_0 = 1$, $\mu_0 = 1$, $\hbar = 1$, $\omega_0 = 5$, $n = 0$, and $s_0 = 1$.

Two kinds of uncertainty products relevant to the Husimi distribution function are considered: one is the usual uncertainty product σ_μ and the other is the more generalized product Σ_μ defined in Eq. (53). While σ_μ varies as time goes by Σ_μ is constant and both have particular relations with those in standard thermal state.

Fock state representation of the Shannon entropies in q- and p-spaces are derived and given in Eqs. (58) and (59), respectively. The EUR which is an alternative uncertainty relation is obtained by adding these two entropies. The EUR is more advantageous than the HUR in the context of information theory. The information theory is not only important in modern technology of quantum computing, cryptography, and communication, its area is now extended to a wide range of emerging fields that require rigorous data analysis like neural systems and human brain. Further developments of theoretical and physical backgrounds for analyzing statistical data obtained from a measurement beyond standard formulation are necessary in order to promote the advance of such relevant sciences and technologies.

Author details

Jeong Ryeol Choi

Department of Radiologic Technology, Daegu Health College, Yeongsong-ro 15, Buk-gu, Daegu 702-722, Republic of Korea

References

[1] Abe, S.; Martinez, S.; Pennini, F. & Plastino, A. (2002). The EUR for power-law wave packets. *Phys. Lett. A*, Vol. 295, Nos. 2-3, pp. 74-77.

[2] Anderson, A. & Halliwell, J. J. (1993). Information-theoretic measure of uncertainty due to quantum and thermal fluctuations. *Phys. Rev. D*, Vol. 48, No. 6, pp. 2753-2765.

[3] Angelow, A. K. & Trifonov, D. A. (2010). Dynamical invariants and Robertson-Schrödinger correlated states of electromagnetic field in nonstationary linear media. arXiv:quant-ph/1009.5593v1.

[4] Ash, R. B. (1990). *Information Theory*. Dover Publications, New York, USA.

[5] Bialynicki-Birula, I. (1984). Entropic uncertainty relations in quantum mechanics. In: L. Accardi and W. von Waldenfels (Editors), *Quantum Probability and Applications*, Lecture Notes in Mathematics 1136, Springer, Berlin.

[6] Biatynicki-Birula, I. & Mycielski, J. (1975). Uncertainty relations for information entropy in wave mechanics. *Commun. Math. Phys.* Vol. 44, No. 2, pp. 129-132.

[7] Bohr, N. (1928). Como Lectures. In: J. Kalckar (Editor), *Niels Bohr Collected Works*, Vol. 6, North-Holand Publishing, Amsterdam, 1985.

[8] Choi, J. R. (2004). Coherent states of general time-dependent harmonic oscillator. *Pramana-J. Phys.*, Vol. 62, No. 1, pp. 13-29.

[9] Choi, J. R. (2010a). Interpreting quantum states of electromagnetic field in time-dependent linear media. *Phys. Rev. A*, Vol. 82, No. 5, pp. 055803(1-4).

[10] Choi, J. R. (2010b). Thermofield dynamics for two-dimensional dissipative mesoscopic circuit coupled to a power source. *J. Phys. Soc. Japan*, Vol. 79, No. 4, pp. 044402(1-6).

[11] Choi, J, R, (2012). Nonclassical properties of superpositions of coherent and squeezed states for electromagnetic fields in time-varying media. In: S. Lyagushyn (Editor), *Quantum Optics and Laser Experiments*, pp. 25-48, InTech, Rijeka.

[12] Choi, J. R.; Kim, M.-S.; Kim, D.; Maamache, M.; Menouar, S. & Nahm, I. H. (2011). Information theories for time-dependent harmonic oscillator. *Ann. Phys.*(N.Y.), Vol. 326, No. 6, pp. 1381-1393.

[13] Choi, J. R. & Yeon, K. H. (2005). Quantum properties of light in linear media with time-dependent parameters by Lewis-Riesenfeld invariant operator method. *Int. J. Mod. Phys. B*, Vol. 19, No. 14, pp. 2213-2224.

[14] Choi, J. R.; Yeon, K. H.; Nahm, I. H.; Kweon, M. J.; Maamache, M. & Menouar, S. (2012). Wigner distribution function and nonclassical properties of schrodinger cat states for electromagnetic fields in time-varying media. In: N. Mebarki, J. Mimouni, N. Belaloui, and K. Ait Moussa (Editors), *The 8th International Conference on Progress in Theoretical Physics*, AIP Conf. Proc. Vol. 1444, pp. 227-232, American Institute of Physics, New York.

[15] Dehesa, J. S.; Martinez-Finkelshtein, A. & Sanchez-Ruiz, J. (2001). Quantum information entropies and orthogonal polynomials. *J. Comput. Appl. Math.*, Vol. 133, Nos. 1-2, pp. 23-46.

[16] Deutsch, D. (1983). Uncertainty in quantum measurements. *Phys. Rev. Lett.*, Vol. 50, No. 9, pp. 631-633.

[17] Fisher, R. A. (1925). Theory of statistical estimation. *Proc. Cambridge Philos. Soc.*, Vol. 22, No. 5, pp. 700-725.

[18] Haldar, S. K. & Chakrabarti, B. (2012). Dynamical features of quantum information entropy of bosonic cloud in the tight trap. arXiv:cond-mat.quant-gas/1111.6732v5.

[19] Husimi, K. (1940). Some formal properties of the density matrix. *Proc. Phys. Math. Soc. Jpn.*, Vol. 22, No. 4, pp. 264-314.

[20] Leplae, L.; Umezawa, H. & Mancini, F. (1974). Derivation and application of the boson method in superconductivity. *Phys. Rep.*, Vol. 10, No. 4, pp. 151-272.

[21] Lewis, H. R. Jr. (1967). Classical and quantum systems with time-dependent harmonic-oscillator-type Hamiltonians. *Phys. Rev. Lett.*, Vol. 18, No. 13, pp. 510-512.

[22] Lewis, H. R. Jr. & Riesenfeld W. B. (1969). An exact quantum theory of the time-dependent harmonic oscillator and of a charged particle in a time-dependent electromagnetic field. *J. Math. Phys.*, Vol. 10, No. 8, pp. 1458-1473.

[23] Lieb, E. H. (1978). Proof of an entropy conjecture of Wehrl. *Commun. Math. Phys.*, Vol. 62, No. 1, pp. 35-41.

[24] Majernik, V. & Richterek, L. (1997). Entropic uncertainty relations. *Eur. J. Phys.*, Vol. 18, No. 2, pp. 79-89.

[25] Malkin, I. A.; Man'ko, V. I. & Trifonov, D. A. (1970). Coherent states and transition probabilities in a time-dependent electromagnetic field. *Phys. Rev. D*, Vol. 2, No. 8, pp. 1371-1384.

[26] Nasiri, S. (2005). Distribution functions in light of the uncertainty principle. *Iranian J. Sci. & Tech. A*, Vol. 29, No. A2, pp. 259-265.

[27] Nielsen, M. A. & Chuang, I. L. (2000). *Quantum Computation and Quantum Information.* Cambridge University Press, Cambridge, England.

[28] Ozawa, M. (2004). Uncertainty relations for noise and disturbance in generalized quantum measurements. *Ann. Phys.*, Vol. 311, No. 2, pp. 350-416.

[29] Pedrosa, I. A. & Rosas, A. (2009). Electromagnetic field quantization in time-dependent linear media. *Phys. Rev. Lett.*, Vol. 103, No. 1, pp. 010402(1-4).

[30] Pennini, F. & Plastino, A. (2004). Heisenberg-Fisher thermal uncertainty measure. *Phys. Rev. E*, Vol. 69, No. 5, pp. 057101(1-4).

[31] Pennini, F. & Plastino, A. (2007). Localization estimation and global vs. local information measures. *Phys. Lett. A*, Vol. 365, No. 4, pp. 263-267.

[32] Pennini, F.; Plastino, A. R. & Plastino, A. (1998) Renyi entropies and Fisher informations as measures of nonextensivity in a Tsallis setting. *Physica A*, Vol. 258, Nos. 3-4, pp. 446-457.

[33] Plastino, A & Casas, A. (2011). New microscopic connections of thermodynamics. In: M. Tadashi (Editor), *Thermodynamics*, pp. 3-23, InTech, Rijeka.

[34] Romera, E.; Sanchez-Moreno, P. & Dehesa, J. S. (2005). The Fisher information of single-particle systems with a central potential. *Chem. Phys. Lett.*, Vol. 414, No. 4-6, pp. 468-472.

[35] Shannon, C. E. (1948a). A mathematical theory of communication. *Bell Syst. Tech.*, Vol. 27, No. 3, pp. 379-423.

[36] Shannon, C. E. (1948b). A mathematical theory of communication II. *Bell Syst. Tech.*, Vol. 27, No. 4, pp. 623-656.

[37] Wehrl, A. (1979). On the relation between classical and quantum-mechanical entropy. *Rep. Math. Phys.*, Vol. 16, No. 3, pp. 353-358.

[38] Werner, R. F. (2004). The uncertainty relation for joint measurement of position and momentum. *Quantum Inform. Comput.*, Vol. 4, Nos. 6-7, pp. 546-562.

Chapter 6

Light Propagation in Optically Dressed Media

A. Raczyński, J. Zaremba and S. Zielińska-Kaniasty

Additional information is available at the end of the chapter

1. Introduction

Light propagation in material media has been the subject of interest for centuries. In particular it has been known that the refraction index, defined as the ratio of the phase velocities in vacuum and in a medium, depends on the light colour, which is known as dispersion. The theory of electromagnetic waves based on Maxwell electrodynamics provided a coherent description of light propagation, which in a material medium is influenced by the medium's polarization. To describe the latter a theory of the atomic medium is required. A simple model of the medium, consisting of classical damped oscillators allowed one to describe the medium's linear response to the propagating field in terms of an electric susceptibility, being in the crudest approximation a Lorentzian function of the light frequency. A causal character of the propagation process implied the analytical properties of the susceptibility. This in turn allowed for drawing important conclusions about the dynamics of propagation and the evolution of the pulse shape, including the presence of the pulse precursors [1].

The birth of quantum mechanics made it possible to describe the atomic structure of the medium in a more sophisticated way. A description of an atom in terms of a wave function, being a superposition of eigenfunctions of a free atom (restricted to a subspace of the states accessible really of virtually due to interaction with light), and of atomic eigenenergies provided a more modern approach to resonant transitions. To account for relaxation, first of all for spontaneous emission, it appeared useful to generalize the quantum formalism by admitting density matrices which after introducing relaxation terms fulfill the optical Bloch equations. The latter equations, completed with the Maxwell equations for the propagating field, treated either classically or quantum-mechanically, constitute a full description of the propagation [2].

A new epoch in the history of studies on light propagation has begun when one realized that by irradiating the medium by an additional (control) field or fields one can completely alter the conditions for the propagation of the probe beam. A striking and important example is the electromagnetically induced transparency (EIT) [3, 4] which consists in making the medium transparent for a pulse resonant with some atomic transition by switching on a strong control laser field, coupling two unpopulated levels. During the last twenty or so years hundreds of papers have been published, studying, both theoretically and experimentally, more and more sophisticated variations of EIT. They include atoms with more active states, coupled by more control fields in various configurations. By admitting the control fields to adiabatically change in time it has become possible to dynamically change the optical properties of the medium while the probe pulse travels inside it [5]. In particular one can reduce the pulse group velocity and finally stop the pulse by switching the control field off; by switching it on again one can release the stored pulse, preserving the phase relations. One can also increase the group velocity or even make it negative. If the control field is in the form of a standing wave, the optical properties of the medium become periodic so the medium resembles a solid state structure [5], which can thus be created on demand with optical means. All those fascinating ways of a precise dynamical control of the optical properties of the medium by optical means reveal new aspects of quantum optics and are supposed to lead to constructing efficient tools for photonics, e.g., quantum memories, quantum switches, multiplexers or, as optimists believe, to designing optically based quantum computers.

Those developments justify the present work the aim of which is to give an introductory review of some of optically dressed atomic systems and to present a method of a theoretical description of their optical properties and of a pulse propagation in such media. In the following chapters we first give a short general theory of wave propagation in atomic media with a few active states. We consider the particular cases of the two-level system, the so-called Λ system, the tripod system and the double Λ system. We show in particular how light propagation and storage can be described in terms of so-called dark state polaritons, being a joint atom+field excitations. We also discuss the situation in which the probe field is described quantum-mechanically which is necessary in the case of a few-photon quantum pulse. In a separate chapter we present atomic models allowing for superluminality, i.e. the pulse's group velocity being negative or larger than the light velocity in vacuum. The final part is devoted to periodic media, a kind of metamaterials, created by optical means. [6].

In our work we remain within the paradigm of the probe field being weak enough to be treated in the linear approximation while the control fields, treated nonperturbatively, are strong enough or couple unpopulated levels so that propagation effects for them can be neglected. We thus leave aside a great part of nonlinear quantum optics dealing with nonlinear effects for the propagating fields.

The bibliography of the field includes hundreds of papers and quickly grows; reviews of various aspects of light propagation in coherently driven atomic media are also available [4, 5, 7, 8]. The list of papers cited here, though obviously not complete, should provide the reader with good tracks for further studies.

2. General theory of light propagation

Consider a quasi-one-dimensional propagation of a probe light pulse in an atomic medium. It is governed by the wave equation stemming from the Maxwell equations

$$\frac{\partial^2 \mathbf{E}_1(z,t)}{\partial z^2} - \frac{1}{c^2}\frac{\partial^2 \mathbf{E}_1(z,t)}{\partial t^2} = \mu_0 \frac{\partial^2 \mathbf{P}_1(z,t)}{\partial t^2}, \tag{1}$$

where $\mathbf{E}_1(z,t) = \mathbf{e}_1 E_1(z,t)$ is the electric field of the probe pulse propagating along the z axis and having polarization $\mathbf{e}_1$, $\mathbf{P}_1(z,t) = \mathbf{e}_1 P_1(z,t)$ is the medium polarization, i.e. the induced dipole moment per unit volume and μ_0 is the vacuum magnetic permeability. Both the electric field and the polarization can be expressed in terms of slowly varying complex amplitudes $\epsilon_1(z,t)$ and $p_1(z,t)$ and a term rapidly oscillating in time and space

$$E_1(z,t) = \epsilon_1(z,t)\exp[i(k_1 z - \omega_1 t)] + c.c.,$$
$$P_1(z,t) = p_1(z,t)\exp[i(k_1 z - \omega_1 t)] + c.c., \tag{2}$$

where ω_1 is the pulse's central frequency and $k_1 = \omega_1/c$; the latter relation, exact for the propagation in vacuum, means that only dilute media are considered, in which the refraction index does not differ much from unity. For not too short pulses one can make the slowly varying envelope approximation (SVEA) [9] which consists in discarding second-order derivatives of ϵ_1 and first- and second-order derivatives of p_1, which leads to the propagation equation of the form

$$\frac{\partial \epsilon_1(z,t)}{\partial t} + c\frac{\partial \epsilon_1(z,t)}{\partial z} = i\frac{\omega_1}{2\epsilon_0} p_1(z,t), \tag{3}$$

where ϵ_0 is the vacuum electric permittivity. If the medium response to the pulse is linear and spatially local, its polarization $p_1(z,t)$ is expressed by a memory time integral

$$p_1(z,t) = \epsilon_0 \int_{-\infty}^{t} \chi(t-t')\epsilon_1(z,t')dt', \tag{4}$$

or, after the Fourier transformation with respect to time,

$$p_1(z,\omega) = \epsilon_0 \chi(\omega)\epsilon_1(z,\omega), \tag{5}$$

where ω is the Fourier variable and χ is the electric susceptibility. The refraction index is $n(\omega) = \sqrt{1+\chi(\omega)} \approx 1 + \frac{1}{2}\chi(\omega)$. Note that the functions in the time or frequency domains are here distinguished only by their arguments. The propagation equation in the Fourier picture takes the form

$$(-i\omega + c\frac{\partial}{\partial z})\epsilon_1(z,\omega) = i\frac{\omega_1}{2}\chi(\omega)\epsilon_1(z,\omega). \tag{6}$$

The last equation can be easily solved and, after returning to the time domain, the solution reads

$$\epsilon_1(z,t) = \frac{1}{2\pi}\int_{-\infty}^{\infty} \epsilon_1(0,\omega)\exp\big(-i\omega t\big)\exp\big[i\frac{\omega z}{c} + i\frac{\omega_1 z}{2c}\chi(\omega)\big]d\omega. \tag{7}$$

In the case of spectrally not too wide pulses one can approximate the susceptibility by the lowest terms of its Taylor expansion at the line centre

$$\chi(\omega) \equiv \chi'(\omega) + i\chi''(\omega) \approx \chi'(0) + i\chi''(0) + \frac{d\chi'(0)}{d\omega}\omega. \tag{8}$$

In such a case one can write the pulse as

$$\epsilon_1(z,t) = \exp\left(i\frac{\omega_1\chi'(0)z}{2c} - \frac{\omega_1\chi''(0)z}{2c}\right)\epsilon_1\left(0, t - \frac{z}{v_g}\right), \tag{9}$$

where the group velocity of the pulse is

$$v_g = c\left(1 + \frac{\omega_1}{2}\frac{d\chi'(0)}{d\omega}\right)^{-1}. \tag{10}$$

This means that the pulse moves with the velocity v_g, with its shape essentially unchanged apart from an exponential modification of its height (Lambert-Beer law) and an overall phase shift. The group velocity is approximately the velocity of the pulse maximum (exactly if there is no damping). A positive value of $\chi''(0)$ corresponds to an exponential damping (absorption) while its negative value - to a negative absorption (gain). Note that derivative of the real part of the susceptibility may be positive (normal dispersion) or negative (anomalous dispersion); in the latter case the group velocity may exceed the light velocity in vacuum or even become negative.

The medium total polarization, due to the total laser field applied to the system, can be expressed in terms of the quantum-mechanical mean value of the dipole moment $\mathbf{d}$

$$\mathbf{P}_1(z,t) = NTr\rho(z,t)\mathbf{d}, \tag{11}$$

where N is the number of atoms per unit volume while $\rho(z,t)$ is the atomic density matrix. It is assumed that atoms, which do not interact with each other, are distributed in a continuous way, with their position along the sample denoted by z.

The atom-field interaction for the atomic system irradiated by the probe field and possibly other control fields is described in the electric dipole approximation, so the hamiltonian for the atom in the position z reads

$$H = H_{at} - \mathbf{E}_1(z,t)\mathbf{d} \tag{12}$$

and the time evolution of ρ is given by the von Neumann equation with some additional phenomenological relaxation terms, known in this context as optical Bloch equation

$$i\hbar\dot{\rho} = [H,\rho] - i\Gamma\rho. \tag{13}$$

The set of Maxwell-Bloch equations provide a complete description of a weak pulse propagation in a dispersive medium. It is assumed that an atom can be represented by a model including a few states $a, b, c, ...$ and each laser field couples a pair of them. Let the probe field 1 be resonant with the transition $a - b$, i.e. it couples the states a (upper) and b (lower) of energies E_a and E_b such that $E_a - E_b \approx \hbar\omega_1$. Due to a lack of resonance or selection rules this field does not couple any other pair of states. Then the part of medium polarization responsible for propagation effects for the field 1 is $P_1 = N(\rho_{ab}d_{ba} + \rho_{ba}d_{ab})$. In the matrix elements of the density matrix one can separate the factor quickly oscillating in time and space, i.e. $\rho_{ab}(z,t) = \sigma_{ab}(z,t)\exp[i(k_1 z - \omega_1 t)]$. A comparison of the terms including the same quickly oscillating factors (the rotating wave approximation - RWA) allows one to write the propagation equation as

$$(\frac{\partial}{\partial t} + c\frac{\partial}{\partial z})\epsilon_1(z,t) = i\frac{N\omega_1 d_{ba}}{2\epsilon_0}\sigma_{ab}(z,t), \tag{14}$$

or in the Fourier picture

$$(-i\omega + c\frac{\partial}{\partial z})\epsilon_1(z,\omega) = i\frac{N\omega_1 d_{ba}}{2\epsilon_0}\sigma_{ab}(z,\omega). \tag{15}$$

Instead of using the field amplitude ϵ_1 one often introduces the so-called Rabi frequency $\Omega_1 \equiv \epsilon_1 d_{ab}/\hbar$, which accounts for the strength of the atom-field coupling for a given transition. The propagation equation Eq. (15) reads then

$$(-i\omega + c\frac{\partial}{\partial z})\Omega_1(z,\omega) = i\frac{N\omega_1 |d_{ba}|^2}{2\epsilon_0 \hbar}\sigma_{ab}(z,\omega) \equiv i\kappa_1^2 \sigma_{ab}(z,\omega). \tag{16}$$

The density matrix element ρ_{ab} or equivalently σ_{ab} (so-called atomic coherence) is obtained from the Bloch equations for a particular atom and coupling model. One need not write down the complete set of those equations due to the assumption of a perturbational treatment of the coupling field E_1. If there were no other sources of this coherence this means that the populations remain unchanged in this approximation and the coherences involving two initially unpopulated states are equal to zero.

3. Light propagation in a few-level atomic media

In this section we review the most important atom-field configurations in the case of which coherent interactions modify the optical properties of an atomic medium. Each of them reveals new physical phenomena and provides one with new means to control the propagation of the probe pulse.

3.1. Two-level atom

In the simplest case of a two-level atom irradiated by the probe field alone [9] (see Figure 1), the only essential equation (i.e. such that it contributes in the first-order perturbation theory with respect to the probe field) is

$$i\hbar\dot{\rho}_{ab}(z,t) = (E_a - E_b - i\hbar\gamma_{ab})\rho_{ab}(z,t) - E_1(z,t)d_{ab}(\rho_{bb} - \rho_{aa}), \tag{17}$$

where ρ_{aa} and ρ_{bb} are initial (and unchanging) populations of the excited state (a) and ground state (b), respectively, and γ_{ab} is the relaxation rate for the coherence ρ_{ab}. In the simplest case,

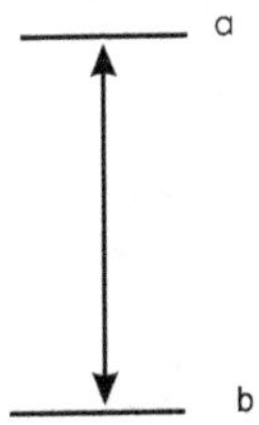

Figure 1. The level and coupling scheme for a two-level system.

in which the spontaneous emission is the only mechanism of relaxation, γ_{ab} is one half of the relaxation rate for the population of the excited state [2].

After separating the rapidly oscillating terms and introducing the Rabi frequency one obtains

$$i\dot{\sigma}_{ab}(z,t) = (-\delta_{ab} - i\gamma_{ab})\sigma_{ab}(z,t) - \Omega_1(\sigma_{bb} - \sigma_{aa}), \tag{18}$$

where $\delta_{ab} \equiv (E_b + \hbar\omega_1 - E_a)/\hbar$ is the laser field detuning, $\sigma_{aa} \equiv \rho_{aa}$ and $\sigma_{bb} \equiv \rho_{bb}$. After the Fourier transformation one obtains the coherence in the frequency domain (we assume that there is no contribution to σ_{ab} other than that due to the probe field)

$$\sigma_{ab}(z,\omega) = -\frac{\Omega_1(z,\omega)}{\omega + \delta_{ab} + i\gamma_{ab}}(\sigma_{bb} - \sigma_{aa}). \tag{19}$$

Thus the propagation equation reads

$$(-i\omega + c\frac{\partial}{\partial z})\Omega_1(z,\omega) = \frac{i\omega_1}{2}\chi(\omega)\Omega_1(z,\omega), \tag{20}$$

with the elelectric susceptibility given by

$$\chi(\omega) = -\frac{N|d_{ab}|^2}{\epsilon_0\hbar}\frac{1}{\omega + \delta_{ab} + i\gamma_{ab}}(\sigma_{bb} - \sigma_{aa}). \tag{21}$$

Note that in the typical case of an unprepared medium, i.e. one being intially in the ground state, $\sigma_{bb} = 1$, $\sigma_{aa} = 0$. Then one has to do with absorption and anomalous dispersion. A typical susceptibility is shown in Figure 2; this plot as well as the following plots of the electric susceptibility are shown for typical values of atomic parameters. The group velocity calculated according to Eq. (10) is larger than c or even negative but absorption at the line center is so strong that the pulse is absorbed just after it has entered the medium and its peak does not even travel inside it. If one has the population inversion, i.e. $\sigma_{aa} > \sigma_{bb}$ the pulse becomes amplified during its propagation. However, one must remember that after it has become strong enough the populations are changed until saturation occurs; such nonperturbative effects are not taken into account in this work.

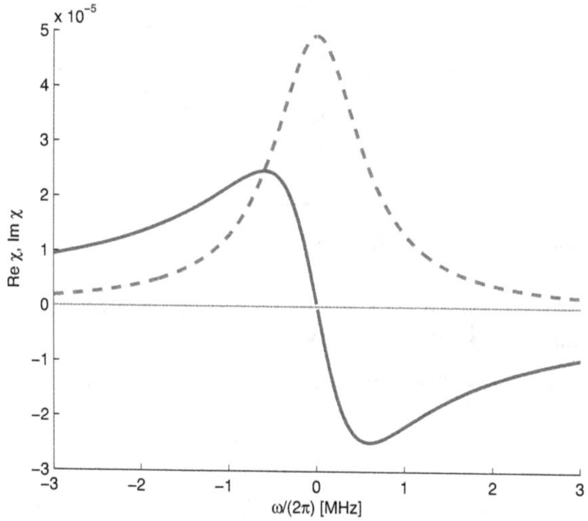

Figure 2. The real (solid, blue line) and imaginary (dashed, red line) parts of the electric susceptibility for a two-level system.

3.2. Λ configuration

Compared with the two-level system, the Λ system includes one additional long-living lower state c. The states b and c may for example be hyperfine or Zeeman states in the lowest atomic electron state. An additional strong laser field E_2 (control field) couples resonantly or almost resonantly the unpopulated states a and c [2]. The level and coupling scheme is shown in Figure 3.

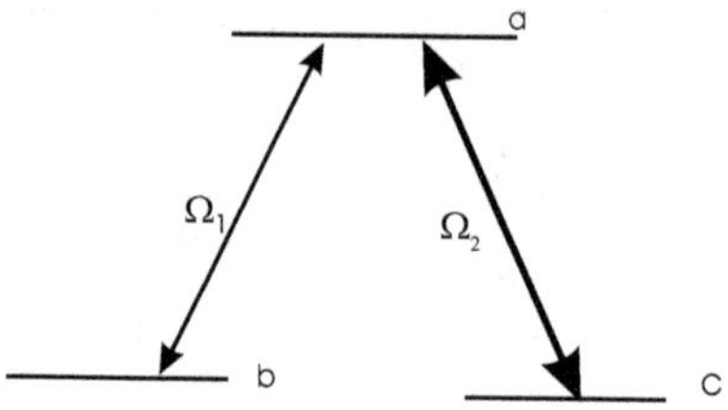

Figure 3. The coupling and level scheme for a Λ system.

The amplitude of the control field is assumed constant both in space and in time. This means that the propagation effects for the control field are neglected; this is justified for a strong control field coupling unpopulated states. The control field is written as

$$E_2(z,t) = E_{20}\exp[i(k_2 z - \omega_2 t)] + c.c. \tag{22}$$

In the case of the only populated state b, $\sigma_{bb}(z,t) = 1$ and the set of essential Bloch equations includes two of them

$$\begin{aligned} i\hbar\dot{\rho}_{ab} &= (E_a - E_b - i\hbar\gamma_{ab})\rho_{ab} - E_1 d_{ab} - E_2 d_{ac}\rho_{cb}, \\ i\hbar\dot{\rho}_{cb} &= (E_c - E_b - i\hbar\gamma_{cb})\rho_{cb} - E_2 d_{ca}\rho_{ab}. \end{aligned} \tag{23}$$

After separating the rapidly oscillating factor, i.e setting $\rho_{cb} = \sigma_{cb}\exp[i((k_1 - k_2)z - (\omega_1 - \omega_2)t)]$ the equations take the form

$$\begin{aligned} i\dot{\sigma}_{ab} &= (-\delta_{ab} - i\gamma_{ab})\sigma_{ab} - \Omega_1 - \Omega_2\sigma_{cb}, \\ i\dot{\sigma}_{cb} &= (-\delta_{ab} + \delta_{ac} - i\gamma_{cb})\sigma_{cb} - \Omega_2^*\sigma_{ab}, \end{aligned} \tag{24}$$

where the Rabi frequency connected with the control field is given by $\Omega_2 = E_{20}d_{ac}/\hbar$, $\delta_{ac} = (E_c + \hbar\omega_2 - E_a)/\hbar$ is the detuning of the control field and γ_{cb} is the relaxation rate for the coherence between the lower states; it is due to collisions between the atoms and collisions of atoms with the walls of the cell and is usually smaller by a few orders of magnitude than γ_{ab}.

In the Fourier picture the equations read

$$\begin{aligned} (\omega + \delta_{ab} + i\gamma_{ab})\sigma_{ab}(z,\omega) &= -\Omega_1(z,\omega) - \Omega_2\sigma_{cb}(z,\omega), \\ (\omega + \delta_{ab} - \delta_{ac} + i\gamma_{cb})\sigma_{cb}(z,\omega) &= -\Omega_2^*\sigma_{cb}(z,\omega), \end{aligned} \tag{25}$$

From the above equations one can calculate the coherence $\sigma_{ab}(z,\omega)$ and then the susceptibility which takes the form

$$\chi(\omega) = -\frac{N|d_{ab}|^2}{\epsilon_0\hbar}\frac{1}{\omega + \delta_{ab} + i\gamma_{ab} - \frac{|\Omega_2|^2}{\omega+\delta_{ab}-\delta_{ac}+i\gamma_{cb}}}. \tag{26}$$

A comparison of the susceptibilities for two-level and Λ systems (see Figures 2 and 4) reveals that switching the control field on leads to producing a dip in the Lorentzian absorption profile, called a transparency window. This means that a resonant probe beam which otherwise would be strongly absorbed, is now transmitted almost without losses. Such a process is known as electromagnetically induced transparency (EIT) [3]. The dispersion inside the transparency window becomes normal, with the slope which increases for a decreasing control field. For a negligible relaxation rate γ_{cb} the absorption dip reaches zero. This means that the medium has become transparent for the probe pulse which travels with a reduced group velocity. The width of the transparency window is proportional to the square of the control field amplitude.

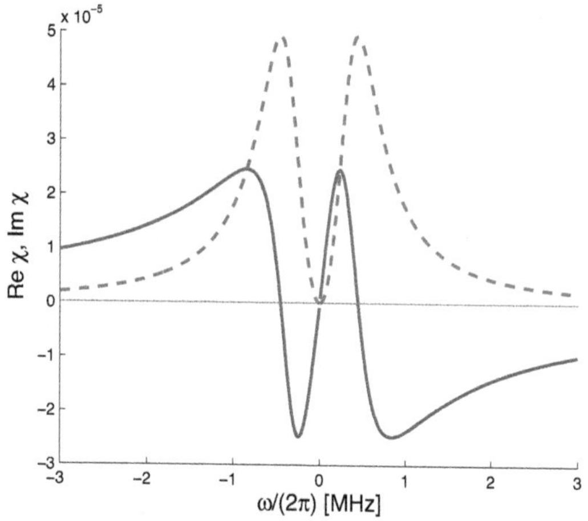

Figure 4. The real (solid, blue line) and imaginary (dashed, red line) parts of the electric susceptibility for a Λ system.

The Maxwell-Bloch equations in the simplest case of resonant ($\delta_{ab} = \delta_{ac} = 0$) and relaxationless conditions ($\gamma_{ab} = \gamma_{cb} = 0$) read

$$\begin{aligned} (\frac{\partial}{\partial t} + c\frac{\partial}{\partial z})\Omega_1(z,t) &= i\kappa_1^2\sigma_{ab}(z,t), \\ i\dot{\sigma}_{ab}(z,t) &= -\Omega_1(z,t) - \Omega_2(t)\sigma_{cb}(z,t), \\ i\dot{\sigma}_{cb}(z,t) &= -\Omega_2(t)\sigma_{ab}(z,t), \end{aligned} \tag{27}$$

where for simplicity Ω_2 is assumed real.

In the adiabatic approximation ($\dot{\sigma}_{ab} = 0$) their solution can be written in terms of the so-called dark state polariton $\Psi(z,t)$ [10] which is a joint atom-field excitation and provides an illustrative insight into the mechanism of propagation

$$\Psi(z,t) = \Omega_1(z,t)\cos\theta(t) - \kappa_1\sigma_{cb}(z,t)\sin\theta(t), \tag{28}$$

where the mixing angle $\theta(t)$ is defined by the relation $\tan\theta(t) = \kappa_1/\Omega_2(t)$. The polariton satisfies the equation

$$(\frac{\partial}{\partial t} + c\cos^2\theta(t)\frac{\partial}{\partial z})\Psi(z,t) = 0, \tag{29}$$

the solution of which is

$$\Psi(z,t) = \Psi(z - c\int_0^t \cos^2\theta(\tau)d\tau, t = 0). \tag{30}$$

The probe field Rabi frequency and the atomic coherence are expressed by the polariton Ψ

$$\begin{aligned} \Omega_1(z,t) &= \Psi(z,t)\cos\theta(t), \\ \sigma_{cb}(z,t) &= -\frac{1}{\kappa_1}\Psi(z,t)\sin\theta(t). \end{aligned} \tag{31}$$

This means that the polariton travels inside the medium without changing its shape, with the velocity dependent on the current value of the control field amplitude. For a decreasing control field the share of the probe field in the polariton decreases and that of the atomic coherence - grows. A gradual switch-off of the control field implies a slowdown of the pulse which can be "stopped" in the limit of $\Omega_2 \to 0$. The width of the transparency window gradually decreases, but the pulse's spectral width is compressed as well [10] so it remains inside the transparency window all the time. Switching the control field on again maps the coherence back into the electromagnetic field, with the phase relations being preserved. This is true provided the relaxation at the storage stage, represented by the relaxation rate γ_{cb}, has not destroyed the phase relations. In practice the times for which a pulse can be stored range from miliseconds in hot gases to seconds in solids. In the language of polaritons light slowdown, stopping and release mean that one changes the ratio of the field and atomic components of the polariton: for a strong control field $\cos\theta$ is almost unity so the polariton is built mainly of the field component; on the opposite, for the control field being switched off $\sin\theta = 1$, which means that the polariton has become purely atomic. One should stress that the expression "light stopping", though illustrative, is a semantic misuse: when the pulse is "stopped", photons constituting the pulse do not exist any more. They have been mapped into an atomic excitation represented by the coherence σ_{cb}; their energy has not been absorbed by atoms but has rather been pumped to the control field. The pulse can be restored by switching the control field on, which provides an inverse mapping of the atomic coherence back into photons. Therefore "light storage" is a more proper term which is commonly used in this context. The successful experiments with stopped light were performed by Liu *et al.* in cold gases [11], Phillips *et al.* in atomic vapours [12] and Turukhin *et al.* in a solid state [13].

Another way of explaining electromagnetically induced transparency applies the notion of dressed states. Consider the subspace spanned by the states a and c (the energy E_c of the latter being moved by the photon energy $\hbar\omega_2$), coupled by the interaction Ω_2. The dressed states are eigenvectors of the hamiltonian restricted to this subspace. The eigenenergies are shifted from their bare values; if the control field is at resonance the shift is equal to $\pm\Omega_2$. If the probe photon's frequency ω_1 is tuned right in the middle between the dressed eigenenergies, the transition amplitudes from the state b interfere destructively. This can be clearly seen if one writes the electric susceptibility for the Λ system in the resonant and relaxationless conditions as

$$\chi(\omega) = -\frac{N|d_{ab}|^2}{2\epsilon_0\hbar}\left(\frac{1}{\omega+\Omega_2}+\frac{1}{\omega-\Omega_2}\right), \tag{32}$$

where the dipole matrix elements between the state b and the dressed states have been expressed by those between b and the bare states (this is where the factor $\frac{1}{2}$ has come from). Indeed, at the centre of the lineshape ($\omega = 0$) the susceptibility takes zero value.

This theory can also be formulated in the fully quantum version [10]. The probe electromagnetic field is quantized

$$\hat{\epsilon}_1(z,t) = \hat{\epsilon}^+(z,t) + \hat{\epsilon}^-(z,t) = \sum_k g_k \hat{a}_k \exp\left(i[(k-k_1)z - (\omega_k - \omega_1)t]\right) + h.c., \tag{33}$$

where $g_k = \sqrt{\frac{\hbar\omega_k}{2\epsilon_0 V}}$, V being the quantization volume, $\hat{a}_k$ and $\hat{a}_k^\dagger$ are photon annihilation and creation operators in the mode k and resonance has been assumed: $E_a - E_b = \hbar\omega_1$. The atomic excitations are also quantized: for an atom in the position z define the flip operator

$$\hat{\sigma}_{bc}(z,t) = |b><c| \exp\left(-i[(k_1-k_2)z - (\omega_1-\omega_2)t]\right), \tag{34}$$

and similarly for other pairs of the indices a, b, c. The time evolution is now governed by the Heisenberg equations of motion. They can be completed by relaxation terms and the corresponding Langevin forces; the latter effect can however be neglected in the timescale of the process [10]; also the relaxation terms will be skipped in the ideal picture presented below. The equations have the same form as Eqs.(24) except that the matrices σ should be transposed. In the adiabatic approximation the solutions can again be written in terms of the polariton field operator

$$\hat{\Psi}(z,t) = \frac{1}{g\sqrt{L}}(\hat{\epsilon}_1^+ \cos\theta - \frac{\hbar\kappa_1}{d_{ab}}\hat{\sigma}_{bc}\sin\theta), \tag{35}$$

where θ is the same mixing angle as before, g is the value of g_k for the central frequency, L is the length of the sample and d_{ab} and Ω_2 are supposed to be real. Note that it was necessary to adopt a different normalization than previously to assure fulfilling the commutation relations typical of creation and annihilation operators

$$[\hat{\Psi}(z,t), \hat{\Psi}^\dagger(z',t] = \delta(z-z'), \tag{36}$$

valid provided the relation $\hat{\sigma}_{bb} \approx 1$ holds, which is true in the first-order of perturbation. The quantum polariton can thus be interpreted as a quasiparticle, being a mixture of the electromagnetic and atomic excitations, the shares of which depend on the current value of the control field. The quantum polariton field operator satisfies the same equation as the classical one (cf. Eq. (29)) and the solution in the adiabatic approximation is

$$\hat{\Psi}(z,t) = \hat{\Psi}(z - c\int_0^t \cos^2\theta(\tau)d\tau, t=0). \tag{37}$$

This means that the evolution of the polariton field operator consists just in changing its position. In particular, switching the control field first off and then on effects in turning the photon into the atomic excitation and back into the photon with certainty and without changing the photon state. Instead of describing creation and annihilation of a quasiparticle at point z one can introduce a family of orthonormal wavepackets $f_j(z)$ and introduce the corresponding operators

$$\hat{\Psi}(j,t) = \int dz f_j^*(z) \hat{\Psi}(z,t). \tag{38}$$

The propagation, storage and release of a single photon accompanied by the atomic excitation characterized by the wave packet f_j are described by the state

$$|1j(t)> = \hat{\Psi}^{\dagger}(j,t)|0>, \tag{39}$$

where $|0>$ is the vacuum state and $|1j(t)>$ is a one-polariton state corresponding to the wave packet f_j. This picture constitutes a basis for possible applications to quantum information processing. An information coded in a single photon state can be written down (stored) as an atomic excitation by switching the control field off and later read out through releasing the photon due to switching the control field on again. In this way one obtains a kind of a quantum memory. The information may be processed inside the memory if the atomic excitation is subject to some controlled transformation during the storage stage. In this particular case of a three-level system it is only the photon phase which can be changed. However, admitting additional active atomic states and additional control fields enables one to perform more sophisticated information processing (see below).

3.3. Tripod configuration

Compared with the Λ configuration, the tripod configuration includes another additional long-living lower state d and a second control field E_3 coupling that state with a (see Figure 5), that is

$$E_3 = E_{30} \exp[i(k_3 z - \omega_3 t)] + c.c., \tag{40}$$

and the corresponding Rabi frequency is $\Omega_3 = E_{30} d_{ad}/\hbar$. Note that it is also possible to consider such a system with two probe fields and one control field or even three fields of comparable intensities but those problems will not be discussed here.

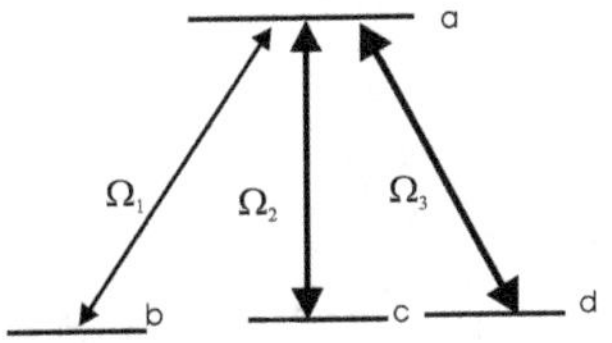

Figure 5. The level and coupling scheme for a tripod system; the field 1 is the probe field, the fields 2 and 3 are control fields.

The essential Bloch equations in the Fourier domain have the form

$$\begin{aligned}
(\omega + \delta_{ab} + i\gamma_{ab})\sigma_{ab}(z,\omega) &= -\Omega_1(z,\omega) - \Omega_2\sigma_{cb}(z,\omega) - \Omega_3\sigma_{db}(z,t), \\
(\omega + \delta_{ab} - \delta_{ac} + i\gamma_{cb})\sigma_{cb}(z,\omega) &= -\Omega_2^*\sigma_{cb}(z,\omega), \\
(\omega + \delta_{ab} - \delta_{ad} + i\gamma_{db})\sigma_{db}(z,\omega) &= -\Omega_3^*\sigma_{db}(z,\omega),
\end{aligned} \tag{41}$$

where σ_{db} is the density matrix element after separating the rapidly oscillating factor, i.e. $\rho_{db} = \sigma_{db} \exp[i((k_1 - k_3)z - (\omega_1 - \omega_3)t)]$, $\delta_{ad} = (E_d + \hbar\omega_3 - E_a)/\hbar$ is the detuning of the second control field and γ_{db} is the relaxation rate for the coherence σ_{db}. The coherence $\sigma_{ab}(z,\omega)$ can be obtained to yield the suscpetibility

$$\chi(\omega) = -\frac{N|d_{ab}|^2}{\epsilon_0 \hbar} \frac{1}{\omega + \delta_{ab} + i\gamma_{ab} - \frac{|\Omega_2|^2}{\omega + \delta_{ab} - \delta_{cb} + i\gamma_{cb}} - \frac{|\Omega_3|^2}{\omega + \delta_{ab} - \delta_{db} + i\gamma_{db}}}. \tag{42}$$

The above susceptibility for the tripod system exhibits in general two transparency windows of different widths and different slopes of the normal dispersion curve (see Figure 6, cf. also Ref. [14]); the latter means that the group velocity in the two windows is different. This asymmetry depends on the model parameters. Speaking in terms of the dressed states one can say that now the subspace in which prediagonalization performed is three dimensional (it is spanned by the states a, c, d), so there are three dressed levels and two frequency regions for which the destructive interference is observed. In general one can analyze configurations including a populated ground state and n unpopulated lower levels, coupled by n control fields with the upper short-living level. The probe field will then experience n transparency windows. If the initial state of the medium is coherently prepared one can observe new effects: two resonant pulses may parametrically generate a third one [15].

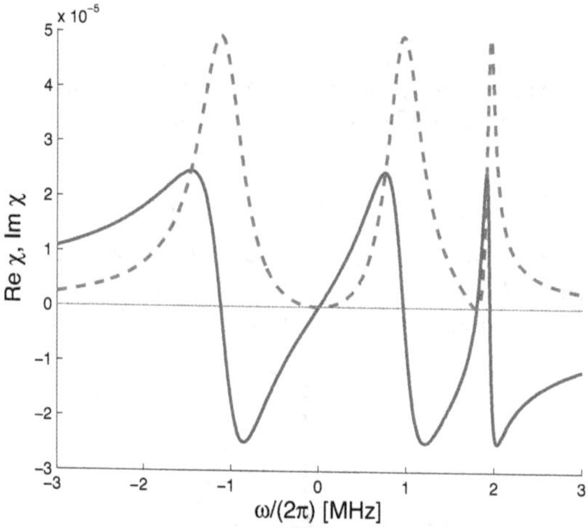

Figure 6. The real (solid, blue line) and imaginary (dashed, red line) parts of the electric susceptibility for a tripod system.

The polariton description in the case of the tripod configuration, both in the classical and quantum versions, is similar to that for the Λ system but requires some modifications [16, 17]. Two polaritons are necessary to describe the adiabatic evolution: they are now built of the electromagnetic component and two atomic excitations. In the quantum version they are represented by the operators

$$\hat{\Psi}_1(z,t) = \frac{1}{g\sqrt{L}}[\hat{e}_1^+ \cos\theta - \frac{\hbar\kappa_1}{d_{ab}}(\hat{\sigma}_{bc}\cos\phi + \hat{\sigma}_{bd}\sin\phi)\sin\theta],$$
$$\hat{\Psi}_2(z,t) = \frac{1}{g\sqrt{L}}\frac{\hbar\kappa_1}{d_{ab}}(\hat{\sigma}_{bc}\sin\phi - \hat{\sigma}_{bd}\cos\phi)\sin\theta, \tag{43}$$

where $\tan\phi = \frac{\Omega_3}{\Omega_2}$ and for simplicity it has been assumed again that the Rabi frequencies of the control fields are real. The equations of motion for the two polaritons read

$$(\frac{\partial}{\partial t} + c\cos^2\theta(t)\frac{\partial}{\partial z})\hat{\Psi}_1(z,t) = \dot{\phi}\hat{\Psi}_2(z,t)\sin\theta,$$
$$\frac{\partial}{\partial t}\hat{\Psi}_2(z,t) = \dot{\phi}\hat{\Psi}_1(z,t)\sin\theta. \tag{44}$$

Thus if the control fields change at the same rate, i.e. $\phi = const$, the polaritons evolve uncoupled; the first one travels with the velocity $c\cos^2\theta$ otherwise unchanged, while the second one is constant in time. Changing the control fields so that $\phi \to \frac{\pi}{2} - \phi$ while $\theta = \frac{\pi}{2}$ means an exchange of the polaritons. In particular the sample may serve as a beam splitter in the time domain [17]: one has to store a single photon in a combination of the atomic excitations using a configuration of the control fields corresponding to some angle $\phi = \phi_0 = const$. If a combination of the control field corresponding to a different angle $\phi = \phi_1 = const$ is applied the photon will be released with the probability $\cos^2(\phi_1 - \phi_0)$. The second part of the release operation corresponding to the angle $\frac{\pi}{2} - \phi_1$ will liberate the photon with probability $\sin^2(\phi_1 - \phi_0)$. The whole operation becomes even more flexible if one admits changing the phases of the control fields. An interesting extrapolation of this idea is a suggestion of a two-photon interference experiment of the Hong-Ou-Mandel [19] type in the time domain. It consists in an independent storing of two photons in two successive steps and in releasing them, also in two steps, using different combinations of the control fields than at the storage stage. For a special combination of the control fields the result is the photon coalescence in one of the two release channels [17], which can be considered an analogue of the Mandel dip in the standard realization. Note that quantum statistical properties, usually concerning photons, can be investigated and modified for the quasiparticles represented by polaritons. In particular, one can store light, perform an operation on the atomic excitations, which changes the statistical properties of the polariton (being a purely atomic excitation at the storage stage), and release light of modified statistical properties.

The picture becomes even more complicated if one applies nonproportional control fields, such that $\dot{\phi} \neq 0$. Figure 7 shows an example the pulse's space-time dependence in the case in which the pulse has been stopped by proportional fields but the releasing field Ω_3 precedes Ω_2. It can be seen that the pulse is released in two stages: in the first one the field Ω_3 liberates the part of the pulse trapped in σ_{db} while in the second one both control fields liberate the pulse from both atomic excitations. Note that the velocities of the two pulse components become equal only after the amplitude of the second control field has reached its final value. A part of the excitation remains in general trapped inside the medium.

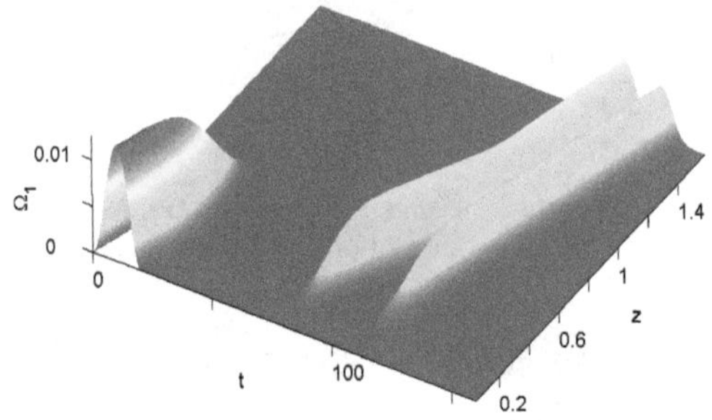

Figure 7. The space-time evolution of the probe field in the case of storing a pulse by proportional control fields and its release by the control fields of the same shape but shifted in time.

Another interesting application was proposed by Wang *at al.* [18], who demonstrated how to obtain a one-photon time-entangled state by storing a single photon and later releasing it in successive steps using different combinations of the control fields. Light storage in a medium of rubidium atoms in the tripod configuration has recently been realized experimentally [20].

3.4. Double Λ configuration

Compared with the single Λ system, the double Λ system considered here includes an additional upper state d coupled with the ground, populated, state b with a second weak probe field $E_3 = \epsilon_3(z,t)\exp[i(k_3 z - \omega_3 t)] + c.c.$, and with the unpopulated state c by a second control field of a constant amplitude $E_4 = E_{40}\exp[i(k_4 z - \omega_4 t)] + c.c.$ (see Figure 8).

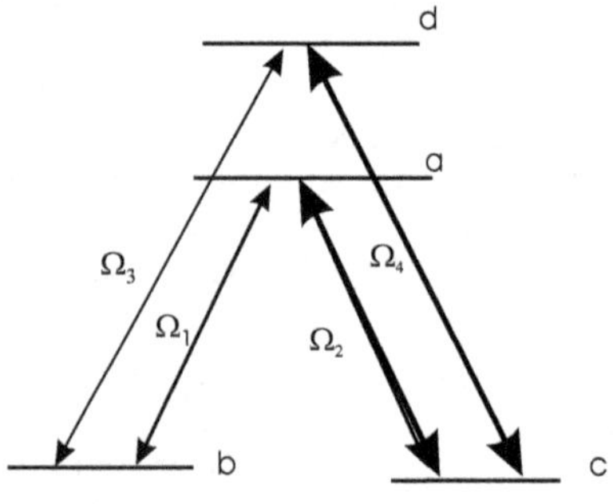

Figure 8. The level and coupling scheme for the double Λ system.

One thus has to do with two Λ's: $b - a - c$ and $b - d - c$. Light propagation in a medium of such a configuration has been investigated in a number of papers [21–25]. It was shown in particular that an adiabatic propagation was possible only for such pulses that $\Omega_1/\Omega_2 = \Omega_3/\Omega_4$. In the general case pulse matching occurs during the first stage of propagation. The essential Bloch equations read

$$\begin{aligned} i\hbar\dot{\rho}_{ab} &= (E_a - E_b - i\hbar\gamma_{ab})\rho_{ab} - E_1 d_{ab} - E_2 d_{ac}\rho_{cb}, \\ i\hbar\dot{\rho}_{db} &= (E_d - E_b - i\hbar\gamma_{db})\rho_{db} - E_3 d_{db} - E_4 d_{dc}\rho_{cb}, \\ i\hbar\dot{\rho}_{cb} &= (E_c - E_b - i\hbar\gamma_{cb})\rho_{cb} - E_2 d_{ca}\rho_{ab} - E_4 d_{cd}\rho_{db}, \end{aligned} \tag{45}$$

where use has been made of the fact that $\rho_{bb} = 1$. Again one can separate the rapidly oscillating factors by substituting additionally $\rho_{db} = \sigma_{db}\exp[i(k_3 z - \omega_3 t)]$, introduce the detuning $(E_b + \hbar\omega_3 - E_d)/\hbar \equiv \delta_{db}$ and the Rabi frequencies $\Omega_3(z,t) \equiv \epsilon_3(z,t)d_{db}/\hbar$ and $\Omega_4 \equiv E_{40}d_{dc}/\hbar$. The above equations take the form

$$\begin{aligned} i\dot{\sigma}_{ab} &= (-\delta_{ab} - i\gamma_{ab})\sigma_{ab} - \Omega_1 - \Omega_2\sigma_{cb}, \\ i\dot{\sigma}_{db} &= (-\delta_{db} - i\gamma_{db})\sigma_{db} - \Omega_3 - \Omega_4\sigma cb\exp(i\phi), \\ i\dot{\sigma}_{cb} &= (-\delta_{ab} + \delta_{ac} - i\gamma_{cb})\sigma_{cb} - \Omega_2^*\sigma_{ab} - \Omega_4^*\sigma_{db}\exp(-i\phi), \end{aligned} \tag{46}$$

where $\phi \equiv (k_1 - k_2 - k_3 + k_4)z - (\omega_1 - \omega_2 - \omega_3 + \omega_4)t$ is a time- and space-dependent phase factor. The analysis of the propagation in the general case of an arbitrary ϕ would require Floquet expansions with respect to the four-wave detuning; in what follows it will be assumed that $\phi = 0$. In this case the equations in the frequency domain read

$$\begin{aligned} (\omega + \delta_{ab} + i\gamma_{ab})\sigma_{ab}(z,\omega) &= -\Omega_1(z,\omega) - \Omega_2\sigma_{cb}(z,\omega), \\ (\omega + \delta_{db} + i\gamma_{db})\sigma_{db}(z,\omega) &= -\Omega_3(z,\omega) - \Omega_4\sigma_{cb}(z,\omega), \\ (\omega + \delta_{ab} - \delta_{ac} + i\gamma_{cb})\sigma_{cb}(z,\omega) &= -\Omega_2^*\sigma_{ab}(z,\omega) - \Omega_4^*\sigma_{db}(z,\omega). \end{aligned} \tag{47}$$

The above equations can be solved with respect to the matrix elements of σ. The solutions can be used in the propagation equations for the two probe fields Ω_1 and Ω_3

$$\begin{aligned} (-i\omega + c\frac{\partial}{\partial z})\Omega_1(z,\omega) &= i\kappa_1^2\sigma_{ab}(z,\omega), \\ (-i\omega + c\frac{\partial}{\partial z})\Omega_3(z,\omega) &= i\kappa_3^2\sigma_{db}(z,\omega), \end{aligned} \tag{48}$$

where $\kappa_1^2 \equiv \frac{N\omega_1|d_{ab}|^2}{2\epsilon_0\hbar}$ and $\kappa_3^2 \equiv \frac{N\omega_3|d_{db}|^2}{2\epsilon_0\hbar}$. One finally obtains coupled propagation equations for the two fields

$$(-i\omega + c\frac{\partial}{\partial z})\Omega_j(z,\omega) = i\sum_{k=1,3} M_{jk}(\omega)\Omega_k(z,\omega), \quad j = 1,3, \tag{49}$$

where

$$
\begin{aligned}
M_{11}(\omega) &= -\frac{\kappa_1^2}{W(\omega)}[(\omega+\delta_{db}+i\gamma_{db})(\omega+\delta_{ab}-\delta_{ac}+i\gamma_{cb})-|\Omega_4|^2], \\
M_{33}(\omega) &= -\frac{\kappa_3^2}{W(\omega)}[(\omega+\delta_{ab}+i\gamma_{ab})(\omega+\delta_{ab}-\delta_{ac}+i\gamma_{cb})-|\Omega_2|^2], \qquad (50)\\
M_{13}(\omega) &= -\frac{\kappa_1^2}{W(\omega)}\Omega_2\Omega_4^*, \\
M_{31}(\omega) &= -\frac{\kappa_3^2}{W(\omega)}\Omega_2^*\Omega_4,
\end{aligned}
$$

$$(51)$$

and

$$
\begin{aligned}
W(\omega) &= (\omega+\delta_{ab}+i\gamma_{ab})(\omega+\delta_{db}+i\gamma_{db})(\omega+\delta_{ab}-\delta_{ac}+i\gamma_{cb}) \\
&\quad -(\omega+\delta_{ab}+i\gamma_{ab})|\Omega_4|^2-(\omega+\delta_{db}+i\gamma_{db})|\Omega_2|^2. \qquad (52)
\end{aligned}
$$

The propagation equations may be decoupled by a linear ω-dependent transformation diagonalizing the matrix M

$$U(\omega)^{-1}M(\omega)U(\omega) = M^d(\omega), \qquad (53)$$

where M^d is the diagonal matrix with

$$M_{11}^d = \frac{1}{2}(M_{11}+M_{33}) + \sqrt{\frac{1}{4}(M_{11}-M_{33})^2 + M_{13}M_{31}}, \qquad (54)$$

$$M_{33}^d = \frac{1}{2}(M_{11}+M_{33}) - \sqrt{\frac{1}{4}(M_{11}-M_{33})^2 + M_{13}M_{31}}. \qquad (55)$$

$$(56)$$

As a consequence, after taking the inverse Fourier transform the solutions $\Omega_j(z,t)$ in the time domain can be written as

$$\Omega_j(z,t) = \frac{1}{2\pi}\int_{-\infty}^{\infty} d\omega \exp\left(-i\omega(t-\frac{z}{c})\right)\sum_{k,m=1,3} U_{jk}(\omega)\exp\left(\frac{iz}{c}(M_{kk}^d(\omega))\right)U_{km}^{-1}(\omega)\Omega_m(z=0,\omega),$$
$$j=1,3. \qquad (57)$$

Thus each of the probe pulses' amplitudes can be considered a superposition of two components, each of which propagates as an analogue of a single pulse with M_{kk}^d playing the role of $\frac{\omega_1}{2}\chi(\omega)$. It appears that one of those "susceptibilities" resembles that for a two-level system while the other one - that for a Λ system. This can be seen from their analytical form in some special cases. For example when $\kappa_1 = \kappa_3$, $\delta_{ab} = \delta_{db}$ and $\gamma_{ab} = \gamma_{db}$ the expressions forM^d_{kk} are identical (apart from the factor ω_1) with the expressions for the susceptibilities given by Eqs. (26) and (21). One can also see that for $\gamma_{cb} = 0$ at the line centre $\omega = 0$ the "susceptibility" $M^d_{33}(\omega = 0) = 0$ so the absorption is zero. This means that one of the superpositions of the pulse amplitudes quickly disappears while the other can propagate unchanged in the conditions of the electromagnetically induced transparency. Disappearance of the former means a transformation of a part of each of the pulses into the other one rather than absorbing the electromagnetic field by the medium. One can also say that pulse matching has occurred which means that the shapes of the two pulses have become adjusted. A description of pulse propagation for such a system can also be formulated in terms of classical [23, 24] or quantum polaritons [25].

4. Superluminal pulses

The notion of the group velocity may still make sense if it happens that its value exceeds that of the light velocity in vacuum or is negative; this depends on the sign and value of the derivative $d\chi'(0)/d\omega$ [1, 26] (see Eq. (10)). Remember that in the case of a two-level atom the absorption is so strong that the group velocity does not correspond to the velocity of the pulse maximum. The situation may be different in the case of the so-called gain doublet [27, 28]. This is a configuration in which there are two closely spaced upper states coupled with the ground state with the probe pulse and the system is prepared so that population inversion occurs.

The electric susceptibility (see Figure 9) resembles then that for the Λ system in the dressed states version

$$\chi(\omega) = \frac{C}{\omega + \delta + i\gamma} + \frac{C}{\omega - \delta + i\gamma}, \tag{58}$$

where, in contradistinction to the case discussed previously, C is now a positive constant and $\omega = 0$ means laser tuning in the middle of the doublet.

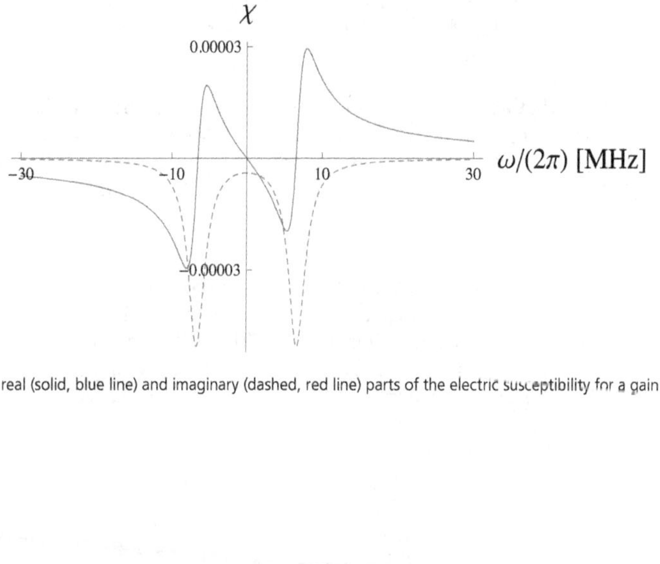

Figure 9. The real (solid, blue line) and imaginary (dashed, red line) parts of the electric susceptibility for a gain doublet.

In this case the dispersion is anomalous but the gain (negative absorption) is not too large. For positive v_g it is almost the velocity of the pulse maximum which moves more quickly than light.

The case of a negative pulse velocity requires more care. It is nonintuitive because the pulse does not any more remain a compact structure. Consider a sample ranging from $z = 0$ to $z = L$. The pulse amplitude should be written as

$$\Omega_1(z,t) = \frac{1}{2\pi}\int_{-\infty}^{\infty} d\omega \Omega_1(z=0,\omega)\exp\left[-i\omega(t-\frac{z}{c}) + i\frac{\omega_1}{2c}\int_0^z \chi(z',\omega)dz'\right] \tag{59}$$

where the susceptibility is constant inside the sample but is zero outside it. Performing the integration yields

$$\Omega_1(z,t) = \frac{1}{2\pi}\int_{-\infty}^{\infty} d\omega \Omega_1(z=0,\omega)\exp\left[-i\omega(t-\frac{z}{c}) + i\frac{\omega_1}{2c}\chi(\omega)\Theta(z)\left(z\Theta(L-z)+L\Theta(z-L)\right)\right], \tag{60}$$

where Θ is the step function. The pulse history is presented in the series of plots in Figure 10 (see also Ref. [29]). When an incoming pulse approaches the entrance of the sample a

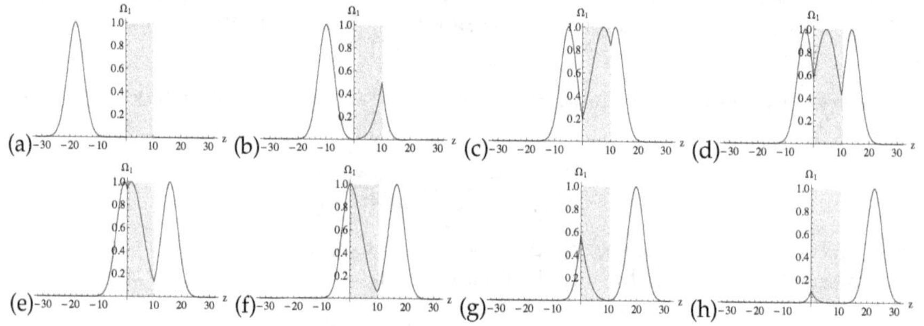

Figure 10. Snapshots of the pulse in arbitrary units, at time instants -18 (a), -10 (b), -5 (c), -3(d), -1 (e), 0 (f), 3 (g), 6 (h); the sample ranges from $z = 0$ to $z = 10$, the pulse velocity in vacuum $v_g = 1$ while in the medium $v_g = -1.5$. The area of the sample has been shaded.

pair of pulses is created at its exit: one inside the sample and the other outside it. The latter moves away from the sample while the former moves backwards with the velocity $v_g < 0$. At the entrance the incoming pulse vanishes and so does that moving backwards inside. The final result is that that the pulse has left the medium earlier than would a pulse traveling in vacuum.

The simplest case in which such effects can in principle be observed is again the Λ system in which however it is the state c which is occupied. Other possible realizations are, e.g., the Λ system with two fields of slightly different frequencies, coupling the the populated state c with a (see the experimental work of Dogariu *et al.* [27]), or the double Λ system with two closely spaced upper levels a and d coupled with the level c by a single control field, with an additional incoherent pump transferring the population from b to c (see Figure 11), or the N-system [30] .

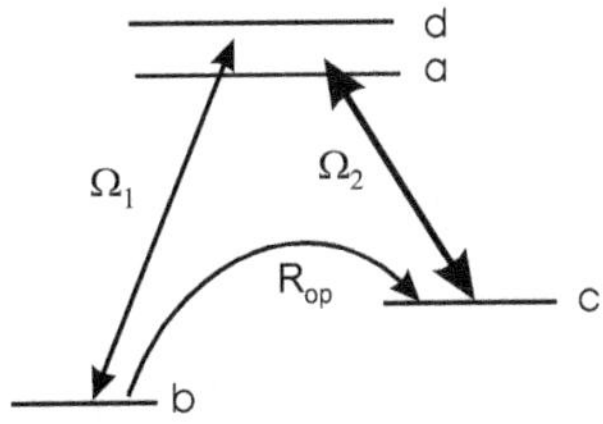

Figure 11. A possible level and coupling scheme for a system with gain doublet. R_{op} is the strength of an incoherent pump.

It is very important to stress that superluminality does not violate causality. It has been shown that neither energy [31] nor information [32] are transferred more quickly than light in vacuum. The proof is based on noticing that the energy of the pulses created at the end of the sample is in some sense "borrowed" from the energy stored inside the medium and not from the energy of the incoming pulse. Another point is that a transfer of information requires a nonanalytical pulse, the spectrum of which is very wide, so it contains a part, built of components of frequencies far from resonance, which propagates unaffected by the medium.

5. Standing-wave control field

New effects occur in a Λ system if the control field is taken in the form of a standing or quasi-standing wave, i.e. one has

$$E_2(z,t) = \left[E_{20+}\exp(ik_2z) + E_{20-}\exp(-ik_2z)\right]\exp(-i\omega_2 t) + c.c., \tag{61}$$

where the subscripts $\pm$ correspond to the direction propagation parallel or antiparallel to the z axis. Such a field makes the optical properties of the medium periodic in space [6, 33, 34]. If the induced lattice fits the incident wave, i.e. $k_1 = k_2$, which also means $\omega_1 = \omega_2$, then in addition to transmission and absorption the incident probe field can be reflected and the methods of describing the propagation are adopted from the solid state physics (Bragg scattering). The probe field including now both the forward and backward propagating components can be written in the two-mode approximation [34]

$$E_1(z,t) = \left[\epsilon_{1+}(z,t)\exp(ik_2z) + \epsilon_{1-}(z,t)\exp(-ik_2z)\right]\exp(-i\omega_1 t) + c.c., \tag{62}$$

where $\epsilon_{1\pm}$ are slowly varying. After considerations similar to those presented in the case of the typical Λ system (the only difference is that now one has to transform off rapid oscillations in time but not in space) one obtains the electric susceptibility rapidly varying in space in the form (cf. eq. (26))

$$\chi(z,\omega) = -\frac{N|d_{ab}|^2}{\epsilon_0\hbar}\frac{1}{\omega+\delta_{ab}+i\gamma_{ab}-\frac{\Omega_{2+}^2+\Omega_{2-}^2+2\Omega_{2+}\Omega_{2-}\cos 2k_2z}{\omega+\delta_{ab}-\delta_{ac}+i\gamma_{cb}}}, \tag{63}$$

where $\Omega_{2\pm} = d_{ac}E_{20\pm}/\hbar$.

The above function can be expanded into the Fourier series

$$\chi(z,\omega) = \chi_0(\omega) + \sum_{j=1}^{\infty} \chi_{2j}(\omega)\big(\exp(2ijk_2z) + \exp(-2ijk_2z)\big). \tag{64}$$

The coupled propagation equations for the two slowly varying components of the probe field read in the frequency domain

$$\begin{aligned} \left(i\frac{\partial}{\partial z} + \frac{\omega}{c} + \frac{\omega_1}{2c}\chi_0(\omega)\right)\Omega_{1+}(z,\omega) + \frac{\omega_1}{2c}\chi_2(\omega)\Omega_{1-}(z,\omega) = 0, \\ \left(-i\frac{\partial}{\partial z} + \frac{\omega}{c} + \frac{\omega_1}{2c}\chi_0(\omega)\right)\Omega_{1-}(z\omega) + \frac{\omega_1}{2c}\chi_2(\omega)\Omega_{1+}(z,\omega) = 0, \end{aligned} \tag{65}$$

where use has been made of the fact that $\omega_1 \approx \omega_2$ and the Rabi frequencies for both components of the probe field have been introduced $\Omega_{1\pm} = d_{ab}\epsilon_{1\pm}/\hbar$. The solutions of the above equations read

$$\begin{aligned} \Omega_{1+}(z,\omega) &= \Omega_{1+}^{+}(\omega)\exp(iQz) + \Omega_{1+}^{-}(\omega)\exp(-iQz), \\ \Omega_{1-}(z,\omega) &= \Omega_{1-}^{+}(\omega)\exp(iQz) + \Omega_{1-}^{-}(\omega)\exp(-iQz), \end{aligned} \tag{66}$$

where the wavevector $Q(\omega)$, the so-called Bloch vector, is given by

$$Q(\omega) = \frac{1}{c}\sqrt{(\omega + \frac{\omega_1}{2}\chi_0(\omega))^2 - (\frac{\omega_1}{2}\chi_2(\omega))^2}, \tag{67}$$

and the superscripts $\pm$ distinguish between the two solutions of the differential equations.

The z-independent functions $\Omega_{1\pm}^{\pm}$ in the above equations should be chosen to guarantee fulfilling the boundary conditions, usually $\Omega_{1+}(0,\omega) = \Omega_{10}(\omega)$, $\Omega_{1-}(L,\omega) = 0$, which corresponds to an incoming wave of the amplitude $\Omega_{10}(\omega)$ entering the sample at $z = 0$ and to the reflected wave equal to zero at its end L. The solutions of Eqs. (65) then read

$$\begin{aligned} \Omega_{1+}(z,\omega) &= \Omega_{10}(\omega)\frac{N_2\exp[iQ(z-L)] - N_1\exp[-iQ(z-L)]}{N_2\exp[-iQL] - N_1\exp[iQL]}, \\ \Omega_{1-}(z,\omega) &= \Omega_{10}(\omega)\frac{\omega_2}{2c}\chi_2(\omega)\frac{-\exp[iQ(z-L)] + \exp[-iQ(z-L)]}{N_2\exp[-iQL] - N_1\exp[iQL]}, \end{aligned} \tag{68}$$

where $N_1 = -Q + \frac{\omega}{c} + \frac{\omega_2}{2c}\chi_0$ and $N_2 = Q + \frac{\omega}{c} + \frac{\omega_2}{2c}\chi_0$. The above expressions can be used to obtain the transmission and reflection spectra for the probe beam

$$T(\omega) = |\frac{\Omega_{1+}(L,\omega)}{\Omega_{10}(\omega)}|^2, \qquad R(\omega) = |\frac{\Omega_{1-}(0,\omega)}{\Omega_{10}(\omega)}|^2, \tag{69}$$

which, together with the dispersion relation $Q = Q(\omega)$, are subject of an experimental verification. A band structure has been created in the medium. In the frequency ranges where $\Re(Q)$ is small and $\Im(Q)$ is considerable, one has to do with band gaps. For incident pulses of such frequencies transmission is forbidden while reflection is strong.

In a more general case, in which the wavenumbers k_1 and k_2 are different, one can make the lattice fit the incident wave by inclining the control beams by an angle β with respect to the z axis so that $k_2' \equiv k_2 \cos\frac{\beta}{2} \approx k_1$ [6]. This means that one should use k_2' instead k_2 in the Fourier expansion of the electric susceptibility and in the two-mode expansion of the probe field. The above formulae take then the form

$$Q(\omega) = \frac{1}{c\cos\frac{\beta}{2}}\sqrt{\left(\omega\frac{\omega_1}{\omega_2} + \frac{\omega_1^2 - \omega_2^2\cos^2\frac{\beta}{2}}{2\omega_2} + \frac{\omega_1^2}{2\omega_2}\chi_0(\omega)\right)^2 - \left(\frac{\omega_1^2}{2\omega_2}\chi_2(\omega)\right)^2}, \tag{70}$$

and

$$N_1 = -Q + \frac{1}{c}\left(\omega\frac{\omega_1}{\omega_2} + \frac{\omega_1^2 - \omega_2^2\cos^2\frac{\beta}{2}}{2\omega_2} + \frac{\omega_1^2}{2\omega_2}\chi_0(\omega)\right),$$

$$N_2 = Q + \frac{1}{c}\left(\omega\frac{\omega_1}{\omega_2} + \frac{\omega_1^2 - \omega_2^2\cos^2\frac{\beta}{2}}{2\omega_2} + \frac{\omega_1^2}{2\omega_2}\chi_0(\omega)\right). \tag{71}$$

A typical dispersion relation $Q(\omega)$ and the corresponding transmission and reflection spectra are shown in Figures 12 and 13. In the interval (stop band) in which the dispersion is almost zero with nonzero absorption the reflection is almost perfect while a narrow transition peak reaches unity where $\Im(Q) \approx 0$. In analogy to the solid state we thus have to do with metamaterials, the optical properties of which can be created on demand with optical methods.

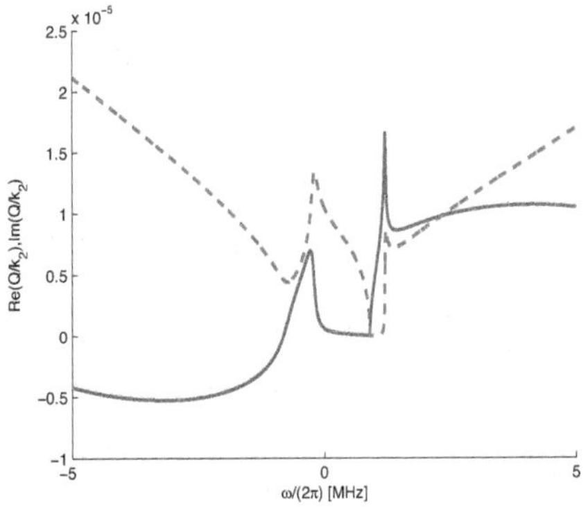

Figure 12. The real (solid, blue line) and imaginary (dashed, red line) parts of the electric susceptibility for a lambda system with control field in the form of quasi-standing wave and a small beam defelction angle β.

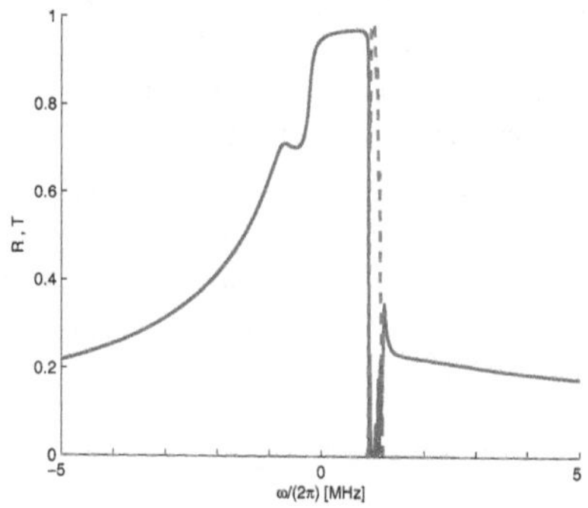

Figure 13. The transmission (dashed, red line) and reflection (solid, blue line) spectra.

6. Conclusions

Various situations - generalizations of electromagnetically induced transparency - have been reviewed in which a weak probe beam of light propagates in an atomic medium which coherently interacts with an additional control field or fields. A unified theoretical description of the particular cases, based on Maxwell-Bloch equations, has been given both for classical and quantum probe fields. The considered cases include in particular EIT, light storage, light processing at the storage stage, pulse matching, superluminality and Bragg scattering on an optically created structure.

Author details

A. Raczyński[1], J. Zaremba[1] and S. Zielińska-Kaniasty[2]

1 Institute of Physics, Faculty of Physics, Astronomy and Informatics, Nicolaus Copernicus University, Toruń, Poland,
2 Institute of Mathematics and Physics, University of Technology and Life Sciences, Bydgoszcz, Poland

References

[1] Jackson J D. Classical Electrodynamics. New York: Wiley; 1975.

[2] Scully M O, Zubairy M S. Quantum Optics. Cambridge: University Press; 1997.

[3] Harris S. Electromagnetically Induced Transparency. Physics Today 1997; 50(7): 36-42.

[4] Fleischhauer M, Imamoğlu A, Marangos J P. Electromagnetically induced transparency: Optics in coherent media. Reviews of Modern Physics 2005; 77: 633-673.

[5] Andre A, Eisaman M D, Walsworth R L, Zibrov A S, Lukin M D. Quantum control of light using electromagnetically induced transparency. Journal of Physics B: Atomic, Molecular and Optical Physics 2005; 38: S589-S604.

[6] Artoni M, la Rocca G C. Optically Tunable Photonic Stop Bands in Homogeneous Absorbing Media. Physical Review Letters 2006; 96(7): 073905 1-4.

[7] Matsko A B, Kocharovskaya O, Rostovtsev Y, Welch G R, Zibrov A S, Scully M O. Slow, Ultraslow, Stored and Frozen Light. Advances in Atomic, Molecular and Optical Physics 2001; 46: 191-242

[8] Milonni P W. Fast Light, Slow Light and Left-Handed Light. Bristol, Philadelphia: Institute of Physcis; 2005.

[9] Allen L, Eberly J H. Optical Resonance and Two-Level Atoms. Dover Publications; 1987.

[10] Fleischhauer M, Lukin M D. Dark-State Polaritons in Electromagnetically Induced Transparency. Physical Review Letters 2000; 84(22): 5094-7.

[11] Liu C, Dutton Z, Behroozi C H, Hau L V. Observation of coherent optical information storage in an atomic medium using halted light pulses. Nature 2001; 409: 490-3.

[12] Phillips D F, Fleischhauer A, Mair A, Walsworth R L, Lukin M D. Storage of Light in Atomic Vapor. Physical Review Letters 2001; 86: 783-6.

[13] Turukhin A V, Sudarshanam V S, Shahriar M S, Musser J A, Ham B S, Hemmer P R. Observation of Ultraslow and Stored Light Pulses in a Solid. Physical Review Letters 2002; 88(2): 023602 1-4.

[14] Paspalakis E, Knight P L. Transparency, slow light and enhanced nonlinear optics in a four-level scheme, Journal of Optics B: Quantum and Semiclassical Optics 2002; 4: S372-5.

[15] Papspalakis E, Kylstra N J, Knight P L. Propagation and nonlinear generation dynamics in a coherently prepared four-level system. Physical Review A 2002; 65: 053808 1-8.

[16] Raczyński A, Rzepecka M, Zaremba J, Zielińska-Kaniasty S. Polariton picture of light propagation and storing in a tripod system. Optics Communications 2006; 260: 73 -80.

[17] Raczyński A, Zaremba J, Zielińska-Kaniasty S. Beam splitting and Hong-Ou-Mandel interference for stored light. Physical Review A 2007; 75: 013810 1-7.

[18] Wang T, Koštrun M, Yelin S F. Multiple beam splitter for single photons. Physical Review A 2004; 70: 053822 1-5.

[19] Hong C K, Ou Z Y, Mandel L. Measurement of subpicosecond time intervals between two photons by interference. Physical Review Letters 1987; 59(18): 2044-6.

[20] Wang H, Li S, Xu Z,Zhao X, Zhang L, Li J, Wu Y, Xie C, Peng K, Xiao M. Quantum interference of stored dual-channel spin-wave excitations in a single tripod system. Physical Review A 2011; 83: 043815 1-6.

[21] Cerboneschi E, Arimondo E. Transparency and dressing for optical pulse pairs through a double-Λ absorbing medium. Physical Review A 1995; 52: R1823-6.

[22] Cerboneschi E, Arimondo E. Propagation and amplitude correlation of pairs of intense pulses interacting with a double-Λ system. Physical Review. A 1996; 54: 5400-9.

[23] Raczyński A and Zaremba J, Controlled light storage in a double lambda system. Optics Communications 2002; 209: 149-54.

[24] Raczyński A, Zaremba J and Zielińska-Kaniasty S. Electromagnetically induced transparency and storing of a pair of pulses of light. Physical Review A 2004; 69: 043801 1-5.

[25] Li Z, Xu L and Wang K. The dark-state polaritons of a double Λ atomic ensemble. Physics Letters A 2005; 346: 269-74.

[26] Milonni P W, Furuya K, and Chiao R Y. Quantum theory of superluminal pulse propagation. Optics Express 2001; 8: 59-65.

[27] Dogariu A, Kuzmich A and Wang L J. Transparent anomalous dispersion and superluminal light-pulse propagation at a negative group velocity. Physical Review A 2001; 63; 053806 1-12.

[28] Steinberg A M, Chiao R Y. Dispersionless highly superluminal propagation in a medium with a gain doublet. Physical Review A 1994; 49: 2071-5.

[29] Ghulghazaryan R, Malakyan Y P. Superluminal optical pulse propagation in nonlinear coherent media. Physical Review A 2003; 67: 063806 1-9.

[30] Kang H, Hernandez G, Zhu Y. Superluminal and slow light propagation in cold atoms. Physical review A 2004; 70: 011801R 1-4.

[31] Diener G. Energy transport in dispersive media and superluminal group velocities. Physics. Letters 1997; 235: 118-24.

[32] Diener G. Superluminal group velocity and information transfer. Physics Letters A 1996; 223: 327-31.

[33] Friedler I, Kurizki G, Pertosyan D. Deterministic quantum logic with photons via optically induced photonic band gaps. Physical Review A 2005; 71: 023803 1-8.

[34] Wu J H, Raczyński A, Zaremba J, Zielińska-Kaniasty S, Artoni M and La Rocca G C. Tunable photonic metamaterials. Journal of Modern Optics 2009; 56(6): 768-83.

Permissions

The contributors of this book come from diverse backgrounds, making this book a truly international effort. This book will bring forth new frontiers with its revolutionizing research information and detailed analysis of the nascent developments around the world.

We would like to thank Andrzej Jamiołkowski, for lending his expertise to make the book truly unique. He has played a crucial role in the development of this book. Without his invaluable contribution this book wouldn't have been possible. He has made vital efforts to compile up to date information on the varied aspects of this subject to make this book a valuable addition to the collection of many professionals and students.

This book was conceptualized with the vision of imparting up-to-date information and advanced data in this field. To ensure the same, a matchless editorial board was set up. Every individual on the board went through rigorous rounds of assessment to prove their worth. After which they invested a large part of their time researching and compiling the most relevant data for our readers. Conferences and sessions were held from time to time between the editorial board and the contributing authors to present the data in the most comprehensible form. The editorial team has worked tirelessly to provide valuable and valid information to help people across the globe.

Every chapter published in this book has been scrutinized by our experts. Their significance has been extensively debated. The topics covered herein carry significant findings which will fuel the growth of the discipline. They may even be implemented as practical applications or may be referred to as a beginning point for another development. Chapters in this book were first published by InTech; hereby published with permission under the Creative Commons Attribution License or equivalent.

The editorial board has been involved in producing this book since its inception. They have spent rigorous hours researching and exploring the diverse topics which have resulted in the successful publishing of this book. They have passed on their knowledge of decades through this book. To expedite this challenging task, the publisher supported the team at every step. A small team of assistant editors was also appointed to further simplify the editing procedure and attain best results for the readers.

Our editorial team has been hand-picked from every corner of the world. Their multi-ethnicity adds dynamic inputs to the discussions which result in innovative outcomes. These outcomes are then further discussed with the researchers and contributors who give their valuable feedback and opinion regarding the same. The feedback is then collaborated with the researches and they are edited in a comprehensive manner to aid the understanding of the subject.

Apart from the editorial board, the designing team has also invested a significant amount of their time in understanding the subject and creating the most relevant covers. They scrutinized every image to scout for the most suitable representation of the subject and create an appropriate cover for the book.

The publishing team has been involved in this book since its early stages. They were actively engaged in every process, be it collecting the data, connecting with the contributors or procuring relevant information. The team has been an ardent support to the editorial, designing and production team. Their endless efforts to recruit the best for this project, has resulted in the accomplishment of this book. They are a veteran in the field of academics and their pool of knowledge is as vast as their experience in printing. Their expertise and guidance has proved useful at every step. Their uncompromising quality standards have made this book an exceptional effort. Their encouragement from time to time has been an inspiration for everyone.

The publisher and the editorial board hope that this book will prove to be a valuable piece of knowledge for researchers, students, practitioners and scholars across the globe.

List of Contributors

Andrzej Jamiołkowski, Miłosz Michalski, Dariusz Chru´sci´nski and Miłosz Michalski
Institute of Physics, Nicolaus Copernicus University, Torun, Poland

Noboru Watanabe
Department of Information Sciences, Tokyo University of Science, Noda City, Chiba, Japan

Jeong Ryeol Choi
Department of Radiologic Technology, Daegu Health College, Yeongsong-ro 15, Buk-gu, Daegu 702-722, Republic of Korea

A. Raczyński and J. Zaremba
Institute of Physics, Faculty of Physics, Astronomy and Informatics, Nicolaus Copernicus University, Toru´n, Poland

S. Zielińska-Kaniasty
Institute of Mathematics and Physics, University of Technology and Life Sciences, Bydgoszcz, Poland